Capturing and Reporting Electronic Data

ACS SYMPOSIUM SERIES **824**

Capturing and Reporting Electronic Data

Willa Garner, Editor
TherImmune Research Corporation

Rodney M. Bennett, Editor
Cerexagri, Inc.

Markus Jensen, Editor
Jensen Agricultural Consultants

American Chemical Society, Washington, DC

Library of Congress Cataloging-in-Publication Data

Capturing and reporting electronic data / Willa Garner, editor, Rodney M. Bennett, editor, Markus Jensen, editor.

p. cm.—(ACS symposium series ; 824)

"Developed from a symposium sponsored gy the Division of Agrochemicals at the 220th National Meeting of the American Chemical Society, Washington, DC, August 20–24, 2000."

Includes bibliographical references and index.

ISBN 0–8412–3784–0

1. Chemical laboratories—Automation—Congresses.

I. Garner, Willa Y., 1936- II. Bennett, Rodney M., 1956- III. Jensen, Markus, 1960 IV. American Chemical Society. Division of Agrochemicals. V. American Chemical Society. Meeting (220th : 2000 : Washington, D.C.) VI. Series.

QD51 .C42   2002
542'.85—dc21   2002018525

The paper used in this publication meets the minimum requirements of American National Standard for Information Sciences—Permanence of Paper for Printed Library Materials, ANSI Z39.48–1984.

Copyright © 2002 American Chemical Society

Distributed by Oxford University Press

All Rights Reserved. Reprographic copying beyond that permitted by Sections 107 or 108 of the U.S. Copyright Act is allowed for internal use only, provided that a per-chapter fee of $22.50 plus $0.75 per page is paid to the Copyright Clearance Center, Inc., 222 Rosewood Drive, Danvers, MA 01923, USA. Republication or reproduction for sale of pages in this book is permitted only under license from ACS. Direct these and other permission requests to ACS Copyright Office, Publications Division, 1155 16th St., N.W., Washington, DC 20036.

The citation of trade names and/or names of manufacturers in this publication is not to be construed as an endorsement or as approval by ACS of the commercial products or services referenced herein; nor should the mere reference herein to any drawing, specification, chemical process, or other data be regarded as a license or as a conveyance of any right or permission to the holder, reader, or any other person or corporation, to manufacture, reproduce, use, or sell any patented invention or copyrighted work that may in any way be related thereto. Registered names, trademarks, etc., used in this publication, even without specific indication thereof, are not to be considered unprotected by law.

PRINTED IN THE UNITED STATES OF AMERICA

Foreword

The ACS Symposium Series was first published in 1974 to provide a mechanism for publishing symposia quickly in book form. The purpose of the series is to publish timely, comprehensive books developed from ACS sponsored symposia based on current scientific research. Occasionally, books are developed from symposia sponsored by other organizations when the topic is of keen interest to the chemistry audience.

Before agreeing to publish a book, the proposed table of contents is reviewed for appropriate and comprehensive coverage and for interest to the audience. Some papers may be excluded to better focus the book; others may be added to provide comprehensiveness. When appropriate, overview or introductory chapters are added. Drafts of chapters are peer-reviewed prior to final acceptance or rejection, and manuscripts are prepared in camera-ready format.

As a rule, only original research papers and original review papers are included in the volumes. Verbatim reproductions of previously published papers are not accepted.

ACS Books Department

Contents

Preface..xi

Acronyms..xiii

1. Introduction to Capturing and Reporting Data
   in the Electronic Age..1
   Kendy L. Keatley

2. Managing Electronic Standard Operating Procedures,
   Protocols, and Protocol Amendments to Good Laboratory
   Practice Requirements...5
   Jane E. Goeke

3. Combination of Old-Fashioned Values and Electronic
   Reporting Systems: Good Laboratory Practice Trial
   Reporting Marriage for the New Millennium..........................13
   W. H. Palmer

4. Development of Advantage™ eFTN: A Good Laboratory
   Practice Field Data Capture System...18
   Ron Thompson, Ted Paczek, and Dudley Dabbs

5. Building the Agrochemical Regulatory Compliance
   Pipleine with Astrix e-Compliance Architecture.....................28
   Richard P. Albert

6. Electronic Field Data Perspective..34
   Kenneth A. Ludwig and Robert Hoag

7. Electronic Data Collection from the Study Director's Viewpoint.......40
   S. Scott Brady

8. Auditing Electronic Field Data 44
    Renée J. Daniel

9. Future of Electronic Good Laboratory Practice Study
    Management Is Now 54
    Fate Thompson, Tommy Willard, and Carla Wells

10. Automated Data Collection in the Laboratory Using
    the Chromeleon™ System 66
    William H. Harned

11. User Testing Strategy for Millennium^{32} Chromatography
    Software Validation 75
    Timothy J. Stachoviak

12. Auditing Electronically Captured Analytical Chemistry Data 79
    Mary E. Lynn

13. Good Laboratory Practice Considerations
    for Electronic Records 86
    Kendy L. Keatley

14. Metrology: A Tool and Approach to Ensure Data Quality 98
    Gail E. Schneiders, John C. Brown, and Joseph Manalo

15. Documentation Requirements for the Design
    of Good Laboratory Practice Software 109
    Robert D. Walla

16. Computer Validation in a Regulatory Environment 117
    Ann M. Speaker and Sharon M. McKilligin

17. Electronic Data Archiving: Ensuring Accessibility,
    Durability, and Usability 124
    Edward J. McDevitt

18. A New Order of Things: Electronic Pesticide Submissions:
    The Promise Is Efficiency 133
    William N. Casey

19. **Benefits of Electronic Data Submissions:
    An Industry Perspective**..139
    Thomas J. Gilding

20. **Food Quality Protection Act of 1996: A New Challenge
    for Data Generation and Submission**..152
    W. T. Beidler and L. D. Bray

21. **Electronic State Submissions**..159
    Charles H. Koopmann

22. **Electronic Pesticide Labeling: Creation, Submission,
    Review, and Dissemination**...164
    Thomas C. Harris

23. **Proposed Revisions to Product Chemistry Data
    Requirements for Registration of Pesticide Chemicals**......................174
    Sami Malak and Deborah McCall

24. **Electronic Data Submission: Pilot Efforts in the Office
    of Pesticide Programs, U.S. Environmental Protection Agency**........188
    Kathryn S. Bouvé

25. **Electronic Data Submission in the Environmental Protection
    Agency Office of Pesticide Programs: Supplemental Files**................198
    Susan V. Hummel

26. **Current State of Electronic Submissions in Europe**..........................209
    Steven C. Dobson

27. **Pest Management Regulatory Agency Experiences:
    Electronic Submissions**...216
    Carmen Krogh

28. **Electronic Data Submissions for the Environmental
    Protection Agency and Pest Management Regulatory
    Agency: Zoxamide Fungicide**..220
    Janet Ollinger, Paul H. Reibach, and Scott Swidersky

29. The Legal and Policy Framework for Electronic Reporting
of Environmental Compliance Reports: Challenges
of E-Government: Maintaining Effective Stewardship
of the Environment..229
M. Evi Huffer

30. Establishing the Groundwork for FIFRA/TSCA Reporting
under Electronic Submissions, Signatures, and Record-Keeping......241
Francisca E. Liem, Mark J. Lehr, and Robert L. Cypher

31. Considerations for Data Collection Systems..253
Rodney M. Bennett

Appendix Establishment of Electronic Reporting: Electronic
Records: Proposed Rule...264

Indexes

Author Index..349

Subject Index...350

Preface

Capturing and reporting data in the electronic age has become a more sophisticated process during the past decade. The magnitude of data that must be collected, correlated, stored and retrieved is continuing to increase exponentially as more and more data are generated on a global landscape. Regulatory oversight of data capturing and handling, and the ability to accurately and quickly store and retrieve these data have become more demanding processes as well. In earlier times, all data were collected, interpreted and stored in some type of paper format. Instrument printouts, as well as calculation sheets, involved recording on a non-electronic medium. Today, most recording devices incorporate some type of electronic capturing equipment.

This book is based upon presentations given during the Division of Agrochemical Symposium and Workshop entitled "Capturing and Reporting Data in the Electronic Age". The symposium and workshop were part of the 220th American Chemical Society (ACS) National Meeting in Washington, D.C., August 20–24, 2000. From the more than 34 papers presented at this symposium, contributing authors prepared the following chapters as a summary.

As we continue to develop more sophisticated, efficient and secure means of data capture, reporting and storage, we must try to keep the process as simple as possible for the end-user. The end-user's are at various stages in the data process and will include those who create the data, those who interpret the data, those who report the data, those who use the data for decision making and those who may need to use the data in the future.

This volume will be of interest to anyone who is involved or anticipates becoming involved in electronic data collection, interpretation or storage. It is a highly useful reference to those individuals who are regulating or are regulated by government or other auditing authorities. The concepts and principals given in this text are applicable to a variety of disciplines within the scientific community. We hope that this volume will provide a basis for additional thought and interaction on the important subject of electronic data for the new millennium.

Acknowledgments

We thank the participants in the symposium and the authors and contributors to this book. Without their interest, enthusiam and efforts, this book would not have been possible. We thank the ACS Division of Agrochemicals for the organizational and financial support it provided for the symposium and shortcourse. In addition, we thank the symposium sponsors for their continuing financial support of the Division of Agrochemicals and all of our technical symposia: The ACS Industry Relations Committee; Cerexagri, Inc. (a wholly-owned subsidiary of ATOFINA Chemicals, Inc.); E.I. DuPont de Nemours; and Syngenta Crop Protection (formerly Novartis Crop Protection).

Willa Garner
Manager, Quality Assurance
TherImmune Research Corporation
620 Professional Drive
Gaithersburg, MD 20879

Rodney M. Bennett
Cerexagri, Inc.
900 First Avenue
King of Prussia, PA 19406

Markus Jensen
Jensen Agricultural Consultants
565 Petite Prairie Road
Washington, LA 70589

Acronyms

AA	Application Administrator
ACES	Access Certificates for Electronic Services
ACPA	American Crop Protection Association
ADEQ	Arizona Department of Environmental Quality
ANDAs	Abbreviated New Drug Applications
ANSI	American National Standards Institute
ASC	Accredited Standards Committee
ATWEED	American Toxicology Work Group on Electronics Exchange of Data
BPPD	Biological and Pollution Prevention Division
CA	Certificate Authority
CADDY	Computer-Aided Dossier and Data Supply
CBER	Center for Biologics Evaluation and Research
CBI	Confidential Business Information
CDER	Center for Drug Evaluation and Research
CDX	Central Data Exchange
CFR	Code of Federal Regulations
cGMP	Current Good Manufacturing Practice
COTS	Commercial Off the Shelf
CRO	Contract Research Organization
CROMERRR	Cross-Media Electronics Reporting and Record-Keeping Rule
CSF	Confidential Statement of Formulas
CWS	Community Water Systems
DACO	Data Code
DAS	Data Acquisition System
DERs	Data Evaluation Records
DMFs	Drug Master Files
ECPA	European Crop Protection Association
EDDE	Electronic Dossier Delivery and Evaluation
EDI	Electronic Data Interchange
eFTN	Electronic Field Trial Notebook

EIEID	Electronics Information Exchange and Interaction Organization
ERR	Electronic Reporting/Record-Keeping
ESTRI	Electronic Standards for the Transmission of Regulatory Information
EU	European Union
FESMCC	Federal Electronic Data Interchange Standards Management Coordinating Committee
FFDCA	Federal Food, Drug, and Cosmetic Act
FI	Field Investigator
FIFRA	Federal Insecticide, Fungicides, and Rodenticide Act
FIPS	Federal Information Processing Standard
FOIA	Freedom of Information Act
FQPA	Food Quality Protection Act
FR	Federal Register
FRFA	Final Regulatory Flexibility Analysis
FTP	File Transfer Protocol
GALP	Good Automated Laboratory Practices
GCPF	Global Crop Protection Federation
GLP	Good Laboratory Practices
GLPS	Good Laboratory Practice Standards
GPEA	Government Paperwork Elimination Act
GRN	Guideline Reference Number
GSA	General Service Administration
HED	Health Effects Division
ICH	International Committee for Harmonization
ICR	Information Collection Request
IEEE	Institute of Electrical and Electronics Engineers, Inc.
IIS	Internet Information Server
INDs	Investigated New Drug Applications
IQ	Installation Qualification
IRFA	Initial Regulatory Flexibility Analysis
ISO	International Organisation of Standardisation
IT	Information Technology
LAN	Local Area Network
LIMS	Laboratory Information Management System
LOD	Level of Detection
LOQ	Limit of Quantitation
MaxLIP	Maximum Likelihood Imputation Procedure
MRID	Master Record Identification Number

MSDS	Materials Safety Data Sheet	
MTS	Microsoft Transaction Server	
NACA	National Agricultural Chemicals Association	
NAFTA	North American Free Trade Agreement	
NARA	National Archives and Records Administration	
NAWQA	National Water Quality Assessment	
NDA	New Drug Application	
NGA	National Governors' Association	
NIST	National Institute of Standards and Technology	
NPDES	National Pollutant Discharge Elimination System	
NTTAA	National Technology Transfer and Advancement Act	
NTWEED	National Toxicology Work Group on Electronics Exchange of Data	
O+M	Operation and Maintenance	
OCR	Optical Character Recognition	
OECD	Organization for Economic Cooperation and Development	
OEI	Office of Environmental Information	
OMB	Office of Management and Budget	
OPMBSTF	Organophosphate Market Basket Survey Task Force	
OPP	Office of Pesticide Programs	
OPPTS	Office of Pollution Prevention and Toxics	
OQ	Operational Qualification	
PAI	Pure Active Ingredient	
PDF	Portable Document Format	
PDP	Pesticide Data Program	
PFI	Principal Field Investigator	
PhRMA	Pharmaceutical Research and Manufacturers of America	
PIN	Personal Identification Number	
PKI	Public Key Infrastructure	
PMA	Project Management Assistance	
PMRA	Pest Management Regulatory Agency	
PQ	Performance Qualification	
PR	Pesticide Registration	
PRA	Paperwork Reduction Act	
PWSS	Pure Water Supply System	
QA	Quality Assurance	
QAP	Quality Assurance Plan	

QAU	Quality Assurance Unit
RAC	Reliability Analysis Center
RED	Reregistration Eligibility Document
RFA	Regulatory Flexibility Act
RTF	Rich Text Format
SDD	Software Design Description
SDLC	Software Development Life Cycle
SEES	State Electronics Commerce/Electronics Data Interchange Steering Committee
SEP	Standard Evaluation Procedures
SOP	Standard Operating Procedures
SRRD	Special Review and Registration Division
SRS	System Requirement Specification
SS	System Suitability
STARR	Sample Tracking and Analytical Results Reporting
TCA	Terms and Conditions Agreement
TCA	Trace Confirmation Analysis
TGAI	Technical Grade Active Ingredient
TNRCC	Texas Natural Resources Conservation Commission
TRI	Toxic Release Inventory
TSCA	Toxic Substances Control Act
TWG	Technical Working Group
UMRA	Unfunded Mandates Reform Act
USDA	U.S. Department of Agriculture
USEPA	U.S. Environmental Protection Agency
USFDA	U.S. Food and Drug Administration
USGS	U.S. Geological Survey
VAN	Value-Added Network
WAP	Wireless Application Protocol

Chapter 1

Introduction to Capturing and Reporting Data in the Electronic Age

Kendy L. Keatley

Gilead Sciences, Inc., 2860 Wilderness Place, Boulder, CO 80301

Both the drug and pesticide industries face new challenges with the regulatory requirements for electronic records and electronic submissions. In addition, the regulations are dynamic with new rules proposed on a regular basis. The use of on-line systems, intranets, and networks allow for distribution of records and documents electronically. Computer systems that create data and generate reports now expand into non-traditional areas, such as, field notebooks and automated weather stations. Topics in this book span the spectrum from electronic data capture systems in the field and in the laboratory to electronic reporting. As always, the bottom line in creating electronic records, in storing or archiving these data, and in submitting data electronically remains the same: the assurance of the integrity and quality of the data.

Computer Validation

The ever-increasing use of electronic systems and automated electronic capture systems provides greater efficiency in the generation and management of records and documents and, thus, replaces paper and manual processes. Where does that electronic system begin? - With the computer. Moreover, all computer hardware dictates the use of its counter part, computer software. All industries under the auspices of the regulatory environment are subject to the requirements of computer validation.

The chapters in this book define the criteria and processes for computer validation. Computer validation applies to all systems, including electronic capture systems in both the laboratory (scientific instrumentation) and field settings or in any other system producing electronic records and documents. The validation process is subject to design specifications, user and performance requirements, preparation of a master plan/validation protocol (installation qualification, operational qualification and performance qualification), execution of the protocol, preparation of a summary report, on-going validation, and re-validation if changes are made. Software life cycle activities extend until retirement of the software. Additional software validation includes implementation of the code, integration, and performance testing. Other predicate rules in place dictate that there is system security, change control procedures, audit trails, calibration, preventative maintenance, and quality assurance.

The validation umbrella not only covers in-house systems, but also covers vendor systems. Much of industry today is dependent to some extent on vendor validation of systems. In light of this, there are validation issues to assess. Vendor considerations include to what extent the vendor plays a role in change control, testing and documentation, source code, integration, and implementation of the system during development, along with implementation support.

Management and Integration of Electronic Records and Documents

It is no longer just data that are produced electronically. A number of documents needed for studies that fall under the Good Laboratory Practice Standards (GLPs) regulations are now being managed electronically. Such records not only include data, such as chromatographic data from automated electronic capture systems and field notebooks, but also documents such as methods, protocols, reports, and Standard Operating Procedures (SOPs). As such, the records are generated, distributed, reviewed, and archived via electronic means.

Compliance with the GLPs for electronic records does not differ from compliance for paper records. However, new challenges are being faced with increased access and distribution of records and documents, and misunderstandings of, for example, the difference between electronic approvals versus electronic signatures. The solution to this challenge lies in assuring system validation and management processes, such as, SOPs or procedures outlining system administration, access, security, change control, training, and disaster planning, are in place.

Electronic data archiving poses more specific challenges. Retention of records and retrieval of records are GLP requirements. Software used for the creation of electronic records not only upgrades at a rapid pace, but also when stored on electronic media, the records still have to comply with the records retention period and be retrievable. The collection, storage, and retrieval of electronic records should address all GLP aspects not to exclude the environmental conditions to ensure the integrity of the media.

With the generation and availability of electronic information also comes the integration of various study information, for example, into a pesticide registration submission, from various locations and sources. Critical issues associated with this process are being challenged with both management of the study and quality assurance. Study Directors now face keeping track of even more data and study reports. Quality assurance departments are challenged with how to audit electronic data. Several chapters in this book provide insight into techniques on how to accomplish these tasks.

Electronic Reporting and Regulatory Submissions

The regulatory world is an exhaustive one, indeed, with new regulations and guidance documents proposed on a regular basis addressing the electronic information world. Although the US Food and Drug Administration (FDA) is already accepting electronic submissions, and the US Environmental Protection Agency has guidance in place for electronic reporting, there is still a lack of a comprehensive regulation or guidance document to establish a standard for reporting. The FDA did issue a final rule on Electronic Records and Electronic Signatures in March of 1997 (*21 CFR Part 11*) (*1*). The rule established regulations applicable for acceptance of records in electronic format and the equivalent of handwritten signatures in electronic form. The rule has, however, raised further questions and the demand for clarity in certain areas with its implementation. Regulators continue to work with one another and with industry to establish a harmonized (international) and comprehensive set of standards.

The Government Paperwork Elimination Act of 1998 requires that all US Federal Agencies provide an electronic reporting platform to the regulated world by the year 2003 (2). There are many benefits to electronic reporting. Electronic reporting can replace paper processes, expedite study reviews, and reduce expenses for both the industry and the Government regarding package preparation and study review, respectively. There are still many issues to address, however. Some of the more important issues are data integrity, confidential business information, the standard platform to use, and how the industry can keep pace with current technologies. A number of chapters address pilots in these areas, new problems that have arisen, and solutions that have been found.

References

1. Food and Drug Administration. Electronic Records; Electronic Signatures; Final Rule. 21 CFR Part 11. 1997; Federal Register, Vol. 62, No. 54.
2. Hogue, C. *C&EN*. September 4, 2000. p.31.

Chapter 2

Managing Electronic Standard Operating Procedures, Protocols, and Protocol Amendments to Good Laboratory Practice Requirements

Jane E. Goeke

Worldwide Regulatory Compliance, GlaxoSmithKline, Mail Stop UE 0376, 709 Swedeland Road, King of Prussia, PA 19406

The use of computer systems to generate, retain, distribute, review and archive standard GLP documentation, such as, SOPs, protocols and protocol amendments, is increasingly common. The benefits include elimination of paper, ease of distribution, increased availability and ease of management. However, failure to consider and/or understand such basic GLP requirements as document availability or where and when electronic signatures are required has resulted in electronic systems that are far less compliant and manageable than simple, old fashioned paper.

There are ten fairly common mistakes in managing electronic SOPs, protocols and protocol amendments. If these can be avoided, the resulting computer system should meet the requirements of both FDA and EPA Good

Laboratory Practices (21 CFR Part 58 and 40 CFR Part 792, respectively) as well as FDA's Rule on Electronic Recordkeeping; Electronic Signatures (21 CFR Part 11) and EPA's proposed rule on Electronic Reporting; Electronic Records.

Mistake #1

Failure to understand that, when using a hybrid system (some combination of paper and electronic records), both mediums are the "real" SOPs, protocols or amendments. As SOP and protocol systems are transferred from a paper to an electronic medium, it is often the practice of maintain both a signed, hard copy and an approved electronic version of the same document.

Users of "hybrid" systems also often forget that it must be possible to demonstrate that the paper and electronic versions of an SOP, protocol or protocol amendment are exactly the same at all times. For example, if the protocol is signed on a certain date, the electronic version must also indicate that approval occurred on the same date. This means that a procedure (SOP) for assuring document equivalency should be available and followed.

Mistake #2

Failure to understand that, if SOPs, protocols and protocol amendments exist only in electronic form, they must be available, especially in the laboratory and study rooms. It is a fundamental FDA and EPA GLP requirement that SOPs be readily available in areas where GLP work is conducted. This means that computer systems which contain these documents must be available in the laboratory and that all personnel who conduct work to these SOPs must have access to the system. The same is true for protocols and protocol amendments.

In cases where a hybrid system exists, i.e., both paper and electronic mediums are simultaneously in use, it is perfectly acceptable that only paper be available in the laboratory area while electronic documents are available in the office. The only caveat here, as noted in Mistake #1, is that a process must be in place to assure the equivalency of the electronic and paper mediums.

One aspect of FDA and EPA GLP requirements that is often overlooked when only electronic systems are used is that, in the event of a system failure, a "back up" paper version is reasonably available. For example, should an electronic SOP system fail, it is unlikely that an FDA/EPA investigator will consider one paper copy of SOPs for a multifloor, multibuilding laboratory as adequate. In GLP terms, the SOPs will not be "readily available".

Mistake #3

Failure to understand that information supporting and tied to electronic SOPs, such as published literature, diagrams or equipment manuals, must also be available and current. It is fairly well understood that, as for paper SOPs, supporting documents such as diagrams or user manuals, should either be incorporated into the SOP or should be "readily available" and current. The same is true for electronic SOPs with the understanding that many current electronic systems are not yet capable of providing such supporting documentation in electronic format. In these cases, it is imperative that the fundamental GLP mandate to somehow tie and reference supporting documentation to the appropriate SOP, as well as to have it be current and available, not be overlooked.

It would also be useful to include such functionality in the user requirements for new or enhanced systems. This is often overlooked.

Mistake #4

Failure to understand that there must be a method to electronically archive historical versions of electronic SOPs as well as study documentation such as protocols and protocol amendments. The FDA is perfectly clear on this requirement in the Electronic Records; Electronic Signatures; Final Rule (21 CFR Part 11). Once an electronic document is created, paper retention is simply not sufficient. It is well understood by both FDA and industry that systems and processes for appropriate electronic archiving may not yet be technically adequate. Despite this, a good faith effort, such as saving historical documents to disk , must be made. Electronic archiving is a function which is often overlooked in developing user requirements.

When hybrid systems are used, it is necessary to retain both electronic and paper versions of SOPs, protocols and protocol amendments since both are "real".

Mistake #5

Failure to understand when electronic signatures are required on electronic records. While all regulated electronic systems must comply with the **electronic recordkeeping** requirements of 21 CFR Part 11, it is not necessary to comply with the **electronic signature requirements** unless signatures are also required by the predicate rule (FDA GLPs, 21 CFR Part 58). In fact, FDA GLPs require signatures only for the following:

- Protocols
- Protocol Amendments
- Reports
- Quality Assurance Reports
- QA Statements
- GLP Compliance Statements
- Raw Data Capture **only when this is performed manually**

It is noteworthy that neither electronic SOPs nor the electronic capture of raw data require electronic signatures. In the case of SOPs, the requirement is that they be "authorized"*(1, 2)* by management. In the case of electronic data, the requirement is that the individual responsible be "identified at the time of data input". *(3, 4)* FDA has made a clear distinction between the terms "sign/initial" versus the terms "approve, authorize, identify". In the first case, an electronic signature is required. In the second case, it is simply necessary that the identity of the responsible individual be clear through use of unique user codes and other means.

What this means in practical terms is that while electronic protocol and protocol amendment systems require an electronic signature, electronic SOPs systems do not. According to FDA's Electronic Records; Electronic Signatures; Final Rule, electronic SOP systems must meet the following requirements:

- system must be validated
- system must provide accurate and complete copies in human readable form
- information must be readily retrievable
- access to system must be limited
- system must have electronic audit trails
- operational, authority and device checks must be part of the system
- system changes occur by change control procedure
- open systems (access not controlled by system owner), require additional security measures such as encryption

Because signatures are required, electronic protocol and amendment systems must meet the following **additional** requirements:

- the signature manifestation must include the printed, full legal name of the signer along with date, time and meaning of the signature
- the signature must be unique to one individual

- the ability to apply the signature must be controlled either by one biometric or two other distinct identification components
- the system must be secure
- device checks must be present
- the company using the signatures must certify to FDA that the electronic signature is the legal equivalent of a written signature

Mistake #6

Failure to validate and manage electronic SOP, protocol and amendment systems to acceptable standards. Validation for such systems must include all standard components including:

- user requirements
- system specifications
- design documentation
- validation plan
- appropriate, well documented testing to include a validation protocol, actual testing (IQ, OQ and PQ), test results, test report and release of the system for use by management.

The validation package should be archived and should be readily retrievable.

After validation is complete, these systems must be used and managed to GLP expectations. At a minimum, this includes:

- operational SOPs including procedures for use, change control and disaster recovery
- limited access including security procedures and a list of authorized users
- periodic testing
- documented staff training
- source code access information
- system overview to include description and diagram

Mistake #7

Failure to consider that electronic SOPs, protocols and amendments must be available to staff at all sites for multisite studies. If the documents will be available at several sites, the validation phase must include functionality testing at each site. Also, documentation of system validation needs to be available at each site as well.

Electronic SOPs should have a limited life span when printed. This is achieved by having each printout state: "Printout not valid after XX date." While this practice assures that system users will not retain printed SOPs long after the electronic SOP is revised, it can present problems when company SOPs are to be followed by an outside contractor which has no access to the electronic system. In such cases, it may be necessary to disable this functionality for the hard copy SOPs which will be provided to the contractor.

Mistake #8

Failure to separate electronic SOPs from electronic policies or guidelines or other documents when the non-SOP documents are managed to a separate standard. While FDA and EPA expect appropriate SOPs for all GLP activities, the agencies have never advised against the use of other documents such as Guidelines or Policies as long as the difference among SOPs, Guidelines and Policies is clear. However, in some organizations, Guidelines and similar documents are used for other purposes such as general reference. In these cases, a decision is often made to manage the Guidelines more loosely than SOPs (e.g., system not fully validated or no operational SOPs exist). In these instances, the regulated, managed SOP system should be fully separated from the other documents and reside on a distinct database.

Mistake #9

Failure to consider that the timing of electronic authorization or electronic signature must meet both FDA and EPA GLP requirements, especially when multiple authorizations or signatures are required. Three examples should suffice to clarify this problem.

Example #1

In a fully electronic SOP system requiring only one manager's "authorization" for new or revised SOPs, the SOPs must not be available on the database before authorization. An electronic SOP must not "go live" on the database until it is approved.

Example #2

In a hybrid protocol system where both the study director and management sign the paper protocol and "authorize" the electronic protocol, the protocol must not be noted as approved on the database without study director signature and authorization, **even though management has signed/authorized**. GLPs require study director signature/authorization for approval. Additionally, since a hybrid system is being used, signing (paper) and authorization (electronic) must occur on the same date.

Example #3

In a fully electronic report system requiring electronic signature by both the study director and management, the report must not appear on the database as finalized **even though management has signed**. Again, GLPs dictate that a report is only final upon application of the study director's signature.

Mistake #10

Failure to create a user friendly system. It is indeed possible to design a computer system for SOPs, protocols and amendments that is fully compliant with FDA and EPA regulations yet so user unfriendly as to be quite burdensome. When designing systems, special attention shall be given to:

- easy access (system and printer availability)
- readable format/easy viewing
- excellent sort capability
- good table of contents
- manageable signature/authorization process
- trackable preparation, review and approval process for new or revised documents
- sensible unique numbering systems
- limitations on number of signatures/authorizations required (no more than two)
- limitations on size of SOPs (not too long)
- a well thought out, well managed transition from paper to electronic documents

References

1. FDA Good Laboratory Practices for Nonclinical Laboratory Studies, 58.81(a), 21 CFR Part 58
2. EPA Good Laboratory Practice Standards, 160.81(a), 40 CFR Part 160
3. FDA Good Laboratory Practices for Nonclinical Laboratory Studies, 58.130(c), 21 CFR Part 58
4. EPA Good Laboratory Practices Standards, 160.130(c), 40 CFR Part 160

Chapter 3

Combination of Old-Fashioned Values and Electronic Reporting Systems: Good Laboratory Practice Trial Reporting Marriage for the New Millennium

W. H. Palmer

A.C.D.S. Research Inc., 4729–A Preemption Road, Lyons, NY 14489

The infidelity of individuals and companies in the 70s submitting data to the Environmental Protection Agency (EPA) led to strict regulations by governmental agencies. In the 90s, the drug and pesticide industries demonstrated the increased efficiency and reduced paperwork achieved with electronic data reporting systems. Governmental agencies responded with more regulations. Electronic reporting systems are now commonplace in both the drug and pesticide industry. However, many electronic data devices are pushed to, and beyond, their environmental limits once these devices are taken out of a laboratory and into the field. In addition, a recent disaster within the industry has demonstrated the vulnerability of magnetic media in comparison to paper records. At the contract research company level, such loss of data or failure of electronic equipment under field conditions could be a catastrophe. A proposal that would match paper record security with the efficiency of electronic data reporting will be presented.

Act (FIFRA) Good Laboratory Practices (GLP) field trials has offered both advantages and disadvantages. This chapter will discuss the author's view on: 1) EPA regulations for electronic data reports and records will be equal to the current FDA regulations for the same; 2) Electronic records offer many advantages for governmental agencies, sponsors, and study directors; 3) Electronic records pose some problems for contract researchers and quality assurance people; and 4) A "marriage proposal" for combination of hard copies and electronic records.

EPA Present and Future Regulations

While the Food and Drug Administration (FDA) has experienced some problems with public approval due to registered products that turned out to have serious safety flaws, the public still is in favor of testing and marketing of new drugs that increase the quality of life. On the other hand, EPA faces a public that mistrusts the pest control industry after the scandals of the 1970s. In addition, the public at large feels that pesticides and genetically modified organisms are unnecessary since we have such an abundance of food in this country. While some people think that EPA will "cut us some slack" for the use of computerized systems in field trials, the author disagrees that EPA will adopt a more lenient attitude. In addition, EPA has stated on several occasions that the agency wants to have consistent regulations, not only between divisions of EPA, but between EPA and other federal agencies. To facilitate this change, EPA has established some new offices and proposed rules including The Office of Environmental Information, the Central Data Exchange, and CROMERRR: Cross-media Electronic Reporting & Record-keeping Rule.

Electronic Record Advantages

Governmental agencies, sponsors, and study directors will benefit from electronic records and reporting systems in the following ways: 1.) Such systems allow compliance with federally-mandated paper reduction standards; 2.) The systems allow faster and easier exchange of information between the sponsors and EPA, between EPA and the public, and between EPA and the states; 3.) The systems will allow more accurate data exchange between these entities; and 4.) The systems will provide uniform types of reports and reporting requirements.

Electronic Records: Potential Problems for Field Trials

Computers for field trials have to perform in an environment that is much harsher than would be experienced in a laboratory situation. EPA's Good Automated Laboratory Practices (GALP) Section 8.7 states that the hardware of computerized systems have to be operated according to the manufacturer's specifications. The temperatures, humidities, static electricity, and dust presence in field trials often exceed the limits of the equipment's specifications. In the proposed combined Toxic Substances Control Act (TSCA) and EPA regulations, computer systems must be validated and there must be plans for security and disaster recovery. The GALP guidelines encourage a "stand-alone system behind locked doors", similar to FDA's "Closed System". FDA's regulations, Section 58.61 call for testing the computerized system in the "target environment".

"Validation" has to be documented evidence that the system will perform consistent, quality records. Even international regulations of the Organization for Economic Co-operation and Development (OECD) defines "acceptance testing" (validation) as formal testing of a computerized system in its anticipated environment. This means that any laptop computer used in field trials would have to be tested under the extremes of temperature, humidity, etc., that might be encountered during a whole testing season. It might be possible to meet these requirements by keeping our computerized systems in a separate, locked, environmentally controlled room with computer access controlled by digital signatures for authorized persons. However, how do we get data collected in the field (calibration data and plot pass times) to the person entering the data into the computerized systems? Cell phones and two-way radios have been used with some success.

In addition, a major fire at a contract facility indicated the vulnerability of magnetic media as opposed to paper records. Even though both types of records had been placed in high-quality fireproof files, only the paper records survived in usable form.

In some instances, the use of an electronic record rather than a paper record limits the ability of the researcher to use time-tested methods of sprayer calibration and other procedures where information is desired. Instead, a "universal" calibration method might be imposed by the trial software that is not the best, or is unfamiliar to the technical staff doing the procedure. The trial software also may not be compatible with facility records and so several pieces of facility data may have to be keyed into the program rather than allowing a simple "cut-n-paste" procedure.

Electronic Records: Problems for Quality Assurance People

An FDA document for its inspectors to use to detect fraud in paper records lists things such as erasure marks, pages that are unusually clean, paper free from stains or smudges (paper should show evidence of normal wear and tear. These easily discerned instances are not available with the pristine electronic records. In addition, 21 CR Part 11 give a perception that Federal agencies may view paper records as being potentially more reliable than electronic records by stating that "paper records are a durable, unitized representation that is fixed in time and space"; paper records are harder to falsify than electronic records, methods to detect falsification of electronic records are yet to be developed, and "there is a perception [in the industry] that electronic records... are less significant.. than paper records."

As A Contractor, Am I Opposed to Electronic Systems?

We have invested over $30,000 to buy software and computerized systems to conduct trials at our three locations in the Northeast. That cost does not include the anticipated larger annual cost for validation, employee training on computerized systems, sofware/hardware upgrades, and more expensive archiving facilities. In our commercial tree-fruit enterprise, we use a sophisticated "smart" airblast sprayer equipped with ultrasonic sensors and computers that allow us to treat only where actual foliage is present. This equipment has allowed us to significantly reduce the amount of pesticides we use on our apples, pears, and stone fruit. Thus, we prefer to use technology rather than vice versa.

A "Marriage Proposal" of Electronic Records and Paper Records

This proposal combines the electronic record advantages of less paper, fast and accurate reporting of field data, a database of information that can be accessed and sorted easily, and uniformity of information forms. For the field researcher, it provides the security of paper records of critical raw data. In any trial, the large majority of information is derived from a relatively small amount of raw data. These raw data, unlike information that can be obtained from alternate sources (i.e., weather data) are critical to reconstruct the trial at a later date. Such critical raw data to be entered on a small number of paper forms include such things as sprayer output, plot length, sprayer speed in the plots, who calibrated equipment and made the application, when they performed those functions, amount of test substance placed in what amount of carrier, etc. These data can be archived at the contract facility as part of the equipment log, transcribed to electronic forms, or sent to the sponsor for archiving.

This system would allow the local researchers to use their familiar standard operating procedures for calibration, etc. If there is a time constraint caused by weather (trying to get applications on before the wind gets too high), then the information can be transcribed to the electronic forms, noted as a transcription, when time permits. This also allows a researcher to use people especially trained for each task, whether it is sprayer operation or data entry. The researcher can then e-mail up-dates of the trial as soon as the data are transcribed. This system avoids the high cost of validation of computerized systems, the use of e-signatures, or the unfortunate loss of critical raw data from a catastrophe such as a fire.

At the end of the trial, another person (maybe from Quality Assurance) checks the electronic file against the raw data preserved by written records. If they are in agreement, then a statement is added to the electronic notebook as to the accuracy of the transcriptions.

For quality assurance people, they will now have the critical raw data on paper so they can use all of the easy methods to check for potential fraud, inaccuracies, etc. In addition, they can now review the paper records at their convenience rather than having to tie up the researcher's computer to check records.

Conclusions

The "marriage" of paper records with electronic records for field trials conducted under FIFRA regulations can be built on the advantages of each data media type. The long-term safety and immediate availability of critical raw data on paper records in ensured. The cost of validation of computerized systems and security is eliminated, thus reducing the cost of trials to the sponsor. Paper records can be used in harshest environmental conditions without compromising recovery of the data. Sponsors have the advantages of fast reporting of field data, easy search and recovery of data that is now in a database, faster report writing, and less archiving of paper records. The combination system would meet the needs of all the stakeholders in the agency/sponsor/contract facility/quality assurance organizations.

Chapter 4

Development of Advantage™ eFTN: A Good Laboratory Practice Field Data Capture System

Ron Thompson, Ted Paczek, and Dudley Dabbs

American Agricultural Services, Inc., 405 East Chatham Street, Cary, NC 27511

Development of Advantage™ electronic Field Trial Notebook (eFTN) as an easy-to-use and cost-effective GLP field data capture system was driven by the need for manipulation of residue data electronically, favorable user economics, ease of use, GLP compliance, and ease of validation. Both companies and regulators need GLP field data in a database that can be queried. Favorable user economics was addressed by Advantage™ eFTN being free to field researchers. Establishing Advantage™ eFTN as a form-driven system and providing flexibility to move Advantage™ eFTN from one computer to another established ease of use. GLP compliance was addressed by requiring unique passwords and time and date stamps for entering the notebook and saving information, the automatic requirement for data revision records for any data changes, and providing multiple security levels. Self-documenting software validation testing insured ease of validation by the field researcher and facilitated compliance acceptance.

Introduction

Just a few years ago, it would have been accurate to say that electronic data capture for GLP (Good Laboratory Practice) field trials will happen. Today, however, a much more accurate statement would be that electronic capture of GLP field data is happening. While less than 5% of GLP field trials in the US in 1998 were conducted using electronic data capture, as noted in the following statement, that had increased substantially by 1999 and continues to increase. "20% of our GLP trials in 1999 were conducted using electronic data capture...we want it, our guys like it, we want it to go that way...it standardizes the notebook (among sponsors) in a way that paper field trial notebooks never accomplished," indicated Mick Qualls, Director of Research, Qualls Agricultural Lab in Ephrata, Washington, at the 1999 meeting of the Society of Quality Assurance.

To understand why this increase in electronic capture of GLP field data occurred so rapidly, it is beneficial to understand the key factors that motivate agrochemical companies to consider electronic study management systems including electronic data capture. This interest in electronics at the sponsor company level is driven primarily by the need for a queryable database and economics. Both EPA and OECD regulators are being inundated with paper to the point that physical storage and handling of paper study reports and submissions is becoming a major burden. In addition, searching and retrieving data from paper systems is inordinately slow and cumbersome. Electronic submissions, which the regulators are openly embracing, are greatly facilitated by having study reports and supporting raw data in an easily accessible format. It is much easier to accomplish this if the studies were managed utilizing electronic study management tools that include electronic data capture and the data are contained in a database that allows the flexibility of electronic reporting.

The second factor that motivates agrochemical companies to consider electronics is simple economics. Agrochemical companies continue to consolidate because the market is maturing such that it can no longer afford a large number of companies and duplication of time and cost. For example, what we recognize today as Aventis Crop Science is composed of a conglomerate of companies that includes what earlier was Hoechst, Schering, Noram, Agrevo, Rhone-Poulenc, Rhodia, Union Carbide, Mobil, Virginia-Carolina, and Amchem. In a similar manner, Syngenta is an aggregation of Novartis and Zeneca each of which, like Aventis, represented quite a list of previous agchem companies that have been consolidated under one management.

In short, the agrochemical industry is asking its internal staff and its subcontractors to continually produce increased results with fewer resources. The challenge to the agchem industry is to find ways to do that easily and cost effectively. Electronic GLP study management and electronic field and laboratory data capture systems are part of the answer.

Advantage™ Field Trial Manager and eFTN Functionality

For purposes of this paper, it is important to understand the principals of electronic GLP study management as discussed in another chapter in this volume entitled "The Future of Electronic GLP Study Management is Now". The functionality of Advantage™ Field Trial Manager and electronic Field Trial Notebook as outlined in Figure 1 and as described in the following text are integral parts of Advantage™ Project Management Assistance (PMA).

Integrated Components

Field trial notebooks that have been generated in Advantage™ PMA are emailed directly from Advantage™ PMA into the second Advantage™ software component, which is Field Trial Manager. Alternatively, Advantage™ electronic Field Trial Notebooks can be mailed to the field researcher by diskette and loaded directly into Advantage™ Field Trial Manager. Sample labels that have been generated in Advantage™ PMA and include critical sample information and bar codes are shipped directly to the field research contractor. Periodically during the field trial, the field researcher can e-mail the Advantage™ eFTNs back to the study director to provide trial updates. At the end of the trial, the completed electronic field trial notebook can be returned to the study director by e-mail or diskette.

Advantage™ Field Trial Manager

The Advantage™ Field Trial Manager software resides at the field researcher's office on either a stand-alone computer or on a server within his local area network. The Field Trial Manager performs several functions for the field researcher including receiving the field trial notebooks from the study director in electronic format and keeping each notebook separate such that it receives and sends out only the data related to that specific field trial and associated electronic Field Trial Notebook.

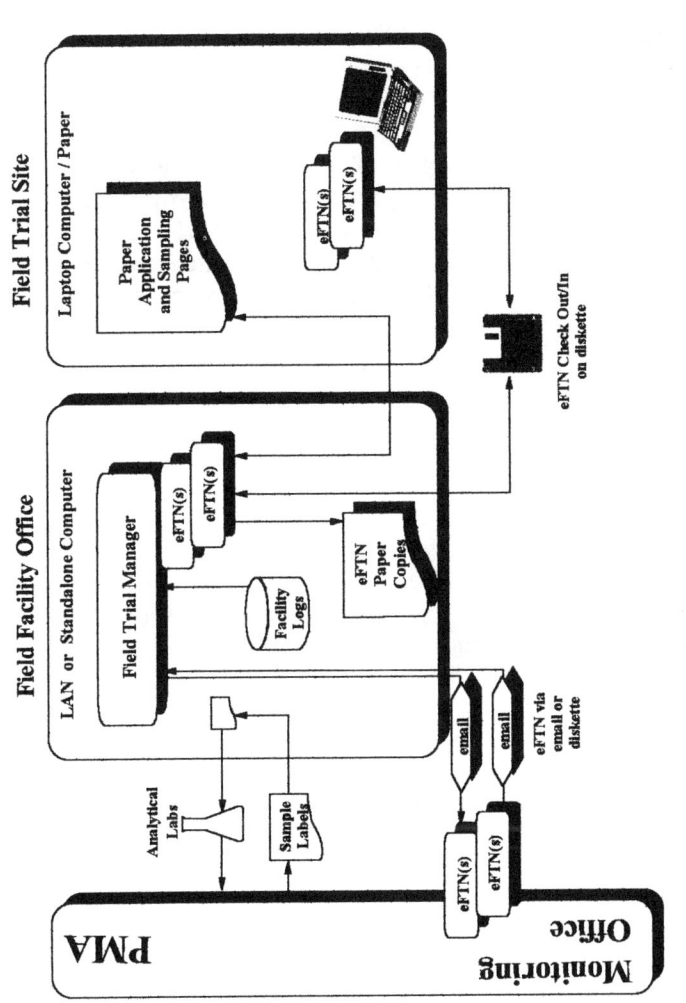

Figure 1. Advantage™ Project Management Assistance and eFTN.

Many field researchers keep their facility records, such as, weather data, field histories, and chemical receipt and storage, using Excel spreadsheets. Advantage™ was intentionally designed so that existing facility records in Excel, etc., formats could be imported directly into Advantage™ in an electronic format. The field researcher does not have to change his current systems or purchase and learn additional software. User experience suggests that the ability to electronically import facility records saves substantial time over paper field trial notebooks. In addition, Advantage™ Field Trial Manager is designed such that the field researcher can draw plot maps in virtually any existing software program and, again, import these directly into Advantage™. This allows the researcher to bring the plot maps into Advantage™ eFTN electronically and avoid the need to learn additional software programs. Typical drawing programs that Advantage™ accommodates include Excel, Power Point, FreeLance Graphics, various CAD systems and Microsoft Paint.

Tremendously flexible print capabilities are incorporated into Field Trial Manager. These enable the field researcher to print any page of the notebook or all pages of the notebook at any time.

eFTN Field Data Collection

Advantage™ eFTN offers by far the most flexibility in the industry in terms of capturing field data. When operating in a completely electronic data capture mode, electronic field trial notebooks can be "checked-out" of Field Trial Manager by copying them to a diskette and, in turn, loading the diskette onto a laptop computer that can be taken directly into the field for direct electronic data entry. Upon completion of the field data collection activity, this process can be reversed by copying the electronic field trial notebook to a diskette and using the "check-in" function to move it back into the Field Trial Manager. Only one "live" copy of the notebook exists at any one time since the notebook 'folder' in Field Trial Manager is locked while it is "checked-out". This prevents the possibility of simultaneous or duplicate entries into eFTN.

Alternatively, most of the eFTN can be completed at the field researcher's office. The print options within Advantage™ Field Trial Manager then allow printing only critical application and sampling pages and taking these pages directly into the field for capture of these critical data on paper. Or, the entire field trial notebook can be printed out and all entries made on paper wherein paper will be the raw data. These data can be subsequently transcribed into the Field Trial Notebook with the retention of paper as the raw data and use of the electronic copy to facilitate data movement.

Development Considerations

Development of Advantage™ electronic Field Trial Notebook as an easy-to-use and cost-effective GLP field data capture system was driven by the need for favorable user economics, ease of use, GLP compliance and ease of validation at the user level. Each of these will be discussed in greater detail below.

Favorable User Economics

Whereas the field research contractor does, and will increasingly in the future, receive some benefits in time and cost savings from using electronic field data capture, the vast majority of the accrued benefits of using electronics occur at the agrochemical sponsor company level. In order for the sponsor companies to receive these benefits and in order for the field research contractor to readily embrace electronic field data capture, American Ag felt that it was imperative that cost to the field researcher for Advantage™ software be kept to a minimum. Therefore, Advantage™ Field Trial Manager and electronic Field Trial Notebook are provided to field research contractors at no cost, thus eliminating cost as a barrier to market entry and user acceptance. As will be noted under "Ease of Use" below, Advantage™ was designed to be intuitive so that little or no training is required thus eliminating time and cost of training as another potential economic hurdle. By contrast, the typical license for each individual computer for FieldNotes™ is $6,000 plus a mandatory annual support fee of $600 per license / computer. In addition, the degree of difficulty in understanding the design and functionality of the FieldNotes software is such that one or more training sessions are required, for which the field researcher must pay a fee plus time and travel.

Ease of Use

American Ag recognized clearly that the field researchers in this industry are excellent field biologists but not computer experts. Therefore, a major development consideration was that Advantage™ eFTN look and feel like a paper field trial notebook. As an indication of our success in accomplishing this, note the following quote from John Corkins, President of Research for Hire in Porterville, California. "Our Principal Field Investigators tell me that if you can use the American Ag paper notebook, you can use the Advantage™

eFTN. Because of this, we don't foresee any additional costs as a result of using Advantage™."

Advantage™ eFTN was designed as a form-driven software program. That is to say, it has the look and feel of a book or of the paper field trial notebook with which field research contractors are already familiar. As the user enters the program, a form appears with sequential tab headings that look like, and function like, the chapters in a book. For example, depending on the trial type, the following headings may appear: General Information, Supporting Forms, Spray Application, Plant Sampling, Storage / Shipping, and Weather Data. At a glance, the user can see the content and functionality of the Advantage™ eFTN chapter after chapter after chapter. Clicking on the chapter heading opens that chapter or section of eFTN and reveals a list of forms or pages with descriptors that reflect the functionality contained on that page. In that regard it looks and feels very much like a book allowing movement easily from chapter to chapter and then page to page within a chapter.

This ability to move chapter to chapter and page to page, either forward or backward, is a reflection of the ease of use and flexibility built into Advantage™ eFTN. The user can always readily see where he is and where he wants to go. Additional flexibility features include the fact that not all information that is requested on a given form has to be included before the user can save existing information and leave that form. Similar flexibility was mentioned earlier with regard to printing wherein any page or all pages of the field trial notebook can be printed at any time prior, during, or after having entered relevant information. Another very beneficial flexibility aspect of Advantage™ is the fact that it can be moved from one computer to another without cumbersome restrictions of licensing being tied to one individual computer as is true for FieldNotes.

A user-friendly aspect of the form driven approach is that a form such as "Spray Equipment Description and Calibration" displays on one page all of the logical information associated with describing the sprayer and necessary settings. Also included is a logical sequence of collection and display of information on calibration to allow automatic electronic calculation of the actual spray rate and volume.

GLP and GALP Compliance

One of the cornerstones of GLP and GALP (Good Automated Laboratory Practice) compliance is that a reviewer be able to recreate what has occurred within a study. More specifically, the reviewer needs to be able to easily determine who did what and when and be able to clearly determine if any changes were made to the original data. Security of the system and the resulting compliance were addressed by requiring a user name and unique password in order to enter Advantage™ Field Trial Manager and, individually, each Advantage™ eFTN. Passwords should be known only to the user and the system administrator and are secure as long as the user and system administrator guard the confidentiality of this information.

Advantage™ contains the capability to assign various security levels for the different individuals who will be using the Field Trial Manager and eFTN. These include a system administrator who has complete access to the system, to add and change user passwords, etc. Additional security levels include "view / edit", which allow field researchers to add, change, and view existing information within the field trial notebook. In addition, "view only" security levels are included for personnel such as quality assurance allowing them to use their user name and password to enter the notebook and view existing data, but does not allow them to add, delete, or change any information.

All activities within the Advantage™ eFTN automatically carry a date and time stamp. Any activity within the Advantage™ eFTN that involves inputting information or changing information automatically inputs a date and time stamp when any information is saved into the system.

A Data Revision Record is provided as part of GLP compliance and to facilitate quality assurance auditing of the field trial notebook. Once data have been entered into the notebook and saved, any attempt to change these data triggers the requirement for a Data Revision Record. If any change of the data is attempted, a form will appear on the screen identifying the data field that is to be changed. It will display the original value and require entry of the new value with an indication of why the data are being changed. In addition, saving this change requires entry of the user name, the unique password, and automatic notation of the date and time at which the data are being changed. This complete Data Revision Record, or any part of it, are available electronically or by printing.

As a further security and compliance measure, Advantage™ contains an automatic proprietary encryption and compression process such that eFTNs sent to or received from field research contractors by either e-mail or diskette can only be read by Advantage™. Included in this process is the ability to prevent field trial data of one sponsor company from being read by another and information in one eFTN from being read into the incorrect file.

Consideration has also been given to archiving requirements for raw data and copies of raw data generated by Advantage™ eFTN. When the field trial is complete, the field researcher uses the "send-to-sponsor" function in Advantage™ Field Trial Manager to email the notebook to the sponsor. This is considered the raw data and can, and should, be archived by the sponsor. In addition, for his own records, the field researcher may also want to archive a copy of the field trial notebook. Both the sponsor and the field researcher can archive electronically in any electronic medium they choose, i.e. hard drive, tape, disc, CD, etc., which prevents deterioration. American Ag maintains software that can read the eFTN regardless of version. Alternatively, the sponsor and field researcher can print the notebook in its entirety, sign, date, and archive it as a hard copy.

User Validation

Software validation can occur at several levels. American Ag has conducted extensive validation of Advantage™ software as the software developer. (See the chapter entitled "The Future of Electronic GLP Study Management is Now".) In addition, sponsor companies who license Advantage™ have validated Advantage™ Project Management Assistance and Advantage™ Field Trial Manger and eFTN as the licensee of this software technology. It can be argued that software can only be validated in the specific environment in which it is being used. Therefore, part of the compliance consideration for Advantage™ Field Trial Manager and eFTN included a comprehensive validation plan containing self-documenting software validation testing to ensure ease of validation by the field researcher.

An easy-to-use validation notebook is provided at no cost to each field researcher that receives Advantage™. This allows him to do a complete user validation on each computer and with each printer, etc. that is utilized in his use environment in collecting and reporting field data. It is comprehensive in that it occurs on the computer system at the point of data collection. Although comprehensive, it is a very easy to use system such that validation of the first computer requires two hours or less and, based on a logical learning curve, requires one hour or less for each additional computer. The validation is comprehensive in that every functional aspect of the notebook is validated according to a validation test script. Documentation is provided to verify, for example, calculations, print integrity, security features, etc. including receipt and return of a field trial notebook sent by e-mail from Advantage™ Project Management Assistance. Reports are then automatically generated from the validation test script to confirm a summary of validation activities and notation of any exceptions that may have occurred.

Conclusion

The need for a queryable database and to obtain increased results with fewer resources has and will continue to drive the acceptance of electronic field data capture. The development of Advantage™ electronic Field Trial Notebook as an easy-to-use and cost-effective GLP field data capture system was implemented and completed with these needs in mind. These aspects plus GLP compliance and ease of validation are the hallmarks of Advantage™ Field Trial Manager and electronic Field Trial Notebook. This software, complemented by Advantage™ Project Management Assistance, comprises the most comprehensive and user-friendly study management software available in the industry today.

Chapter 5

Building the Agrochemical Regulatory Compliance Pipleine with Astrix e-Compliance Architecture

Richard P. Albert

Astrix Software Technology, Inc., 175 May Street, Suite 302, Edison, NJ 08837

> The use of web-centric information technology systems will allow life science companies to bring their products to market faster, more efficiently and, ultimately, more safely. The process of registering a new drug or pesticide with the regulatory authorities is a long and data intensive process with numerous stakeholders. Pharmaceutical and Agrochemical companies are global organizations and the trials necessary to support product registration produce a vast amount of data.

Overview

Each participant throughout the process and across the enterprise has individual informational needs which change on a regular basis and need to be shared and disseminated throughout the organization. Astrix has developed a strategy that allows global life science companies to comply with the regulatory requirements using a web-centric information technology system (e-compliance system).

Characteristics of the Agrochemical Registration Process

The Agrochemical registration process is composed of multiple steps that are interdependent. The integrity and timeliness of the completion of the entire process is dependent on successfully executing each of the steps. The process involves participants that are located at remote sites throughout the world. Many of the participants are third party contractors that also play a vital role in collecting data to support the regulatory approval process. Some common characteristics found in the agrochemical registration process include:

- Deciding on new formulations, expanding product labels, and determining the company sales and marketing initiatives are essential business that have many participants.
- Study worksheets, bid-sheets, and laboratory and field protocols are developed, reviewed and approved by various participants
- The field part of the study is dependent on the agrochemical practices of the remote field site. Therefore, review and input by the remote site is essential.
- The laboratory must have access to methodologies, method validation information, and compound specific information.
- Data from multiple sources are sent to the sponsor from many participants, both internal and external.
- Many people, including field coordinators, lab managers, study directors, quality assurance personnel, farm managers, and others, are involved in the process.
- The generation of quality data that are complete, legible, traceable, and meet the requirements of the regulations is essential.
- The regulatory agency is the ultimate client and the generation of an electronic deliverable with essential information organized for easy evaluation and review is critical.

The agrochemical registration process involves many partners that must work together to reach the ultimate goal of adequately and efficiently reviewing an EPA regulatory submission. The sponsor, field contractors, analytical laboratories, and the regulatory agency are all primary stakeholders in the process (See Figure 1).

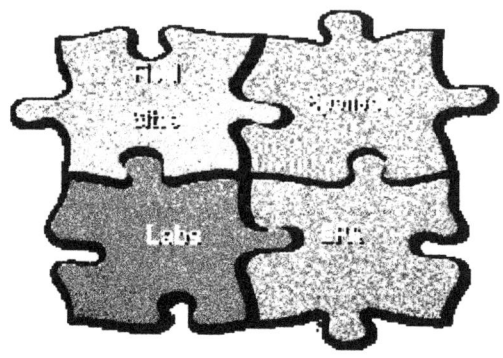

Figure 1. Partner Participation.

Astrix Agrochemical e-Compliance System

Astrix Software Technology is building the next generation agrochemical management system to allow companies to manage the entire regulatory process via the corporate intranet. The system will allow authorized users to collaborate on the development of new ideas, generate worksheets and bid sheets, develop protocols, generate sample labels, web forms for the field, analytical reporting forms, and provide sample tracking, management, and financial data.

The agrochemical e-compliance server is designed to allow multiple people to develop, review, and authorize documentation and data used in the registration process. The system operates by using the corporate intranet/extranet as the communication backbone for the system.

Astrix Agrochemical E-Compliance Architecture

The technical architecture of the agrochemical e-compliance server is built around the premise of flexibility and scalability. The system is designed in three tiers (Figure 2). The components that comprise the architecture include agrochemical e-compliance interface, middle tier Astrix Ag COM components including middle tier reporting objects, and the Astrix Agrochemical database. The system components reside on a Windows NT server:

Presentation Tier (User Interface)

The presentation layer resides on a Windows NT Server running Microsoft Internet Information Server (IIS). All screens (forms) and reports are displayed in a web browser.

Business Logic Tier (Astrix Ag COM Components)

The application layer resides on a Windows NT Server running Microsoft Transaction Server (MTS). All business logic is encapsulated in objects.

Data Tier (Agrochemical Database)

The database layer resides on a Windows NT Server running an installation of Oracle 8 Enterprise Server Database.

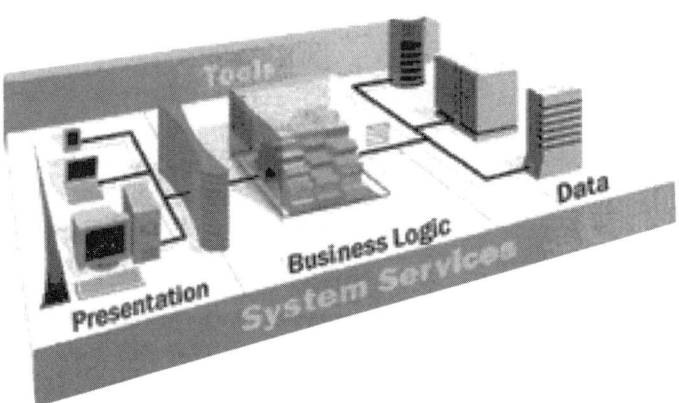

Figure 2. Agrochemical e-Compliance Server Architecture.

The primary advantage of a web-centric architecture is that all business logic, data, and system functionality are maintained centrally on a web server. The system is a "zero footprint" architecture, meaning that no components are installed on the client. This eliminates conflicts with client operating systems, dynamic link library incompatibilities and a host of other problems resulting from software installation. The system can also be updated centrally without users needing to perform an upgrade. The data are stored behind a firewall in a secure Oracle 8 repository. Central data storage is the key security feature of the system.

Astrix e-Compliance Project Planning Module

The regulatory study process originates with a new idea. New ideas can originate from many different sources in the organization and generally consist of new market opportunities for a product, such as, adding a new crop to a label or developing a new formulation. Typically, a preliminary project plan will be developed that includes a project timeline and estimated development costs. The database serves as a repository for vital information that allows the stakeholders to access, mine and evaluate data. A viability review can be conducted on line to evaluate the strategic fit, analyze financial information, and determine what data are needed to support the idea. If the results of the evaluation support the idea, it will move to the Assessment Phase. Input on the proposed project will be requested from the various functional groups to supplement the information from the review phase. A Preliminary Project Plan is developed on line that includes a project timeline and estimated development costs.

Astrix e-Compliance Protocol Module

Once the project has been approved it enters the protocol development process. The first step in the process is the completion of the protocol worksheets. The system allows users to create dynamic worksheets and bid sheets that can be automatically sent to contractors and collaborators. The draft protocols are developed from the information contained in the worksheets. Both the field and laboratory protocols are developed by various participants, usually Study Directors, and can be distributed via the corporate intranet for peer review.

Astrix e-FieldNotes System

The protocol information can be assessed via a web browser by the field contractor. The observational in-life data can be entered on a standard desktop PC directly via an extranet connection to the server. Data that must be collected in the field (i.e., application data) can be collected with a Palm OS based unit (i.e., Palm Pilot) and entered directly in the field. At the end of the day the Palm unit can be synchronized with the web server database. This architecture provides the user with the ultimate mobility and results in a high level of data security. The e-FieldNotes System is designed using the Wireless Application Protocol Technology (WAP) that is optimized for low bandwidth environments that are typically found in rural locations.

Astrix e-Laboratory System

Astrix supports a standard data exchange file format that can be automatically generated either by the instrument data system (i.e., ChemStation, TurboChrom) or by a commercially available LIMS (i.e., PE LIMS) system. In addition, the laboratory can assess web-based forms and manually enter the analytical results directly into the server database. The system also supports custom generated reports and queries.

Astrix e-Report Module

The final report system allows a user located anywhere in the world to access data from the Agrochemical repository database and generate reports. Depending on user access and security levels the user can generate the protocol and field, laboratory or final data reports. In addition to data reports the system generates management summaries including both financial and resource allocation reports.

Summary

Astrix Software Technology is a leading supplier of e-compliance tools to the environmental, pharmaceutical, medical device, and agrochemical industries. Web-centric computer systems allow data to be collected and accessed remotely while maintaining central control of the data and the application components. The corporate intranet brings customers, regulators, collaborators, and employees together. As life science companies continue to consolidate to face the challenges of a global marketplace the information technology providers must respond to this changing landscape. A global strategy for Agrochemical research and development must be supported with a system that extends throughout the enterprise and into every corner of the world.

Chapter 6

Electronic Field Data Perspective

Kenneth A. Ludwig[1,2] and Robert Hoag[2]

[1]Bayer Corporation, Agriculture Division, Bayer Research Park, 17745 South Metcalf, Stilwell, KS 66085–9104
[2]Current Address: 1611 Padock Drive, Kearney, MO 64060–8423

The Agriculture Division of Bayer Corporation has been using the Astrix FieldNotes™ electronic field notebook in field residue studies since 1997. During 1997 it was used for all field residue trials conducted on Bayer research farms. Beginning in 1998, it has been used for all Bayer field residue trials. Bayer uses Astrix LabNotes™ to manage field residue trial samples in the laboratory. In addition, Bayer uses a spreadsheet and word processor to create study protocols, bid forms and work agreements. A spreadsheet is used to facilitate management of the field trials. All of this electronic information is kept on the Bayer network where Bayer employees may readily access it. Information transmission is done on the network within Bayer and via the Internet with contractors. Electronic systems have been important tools that have given Bayer the ability to substantially increase the number of residue studies conducted each year.

Introduction

Electronic capture of field trial data seems to be developing more slowly than electronic capture of laboratory data and field performance data. This may be, in part, because GLP (Good Laboratory Practice) requirements do not apply to performance research. Major differences among performance, magnitude of residue and ecological effects trials, resulting in relatively small markets for electronic applications are probably major factors. This chapter concentrates on the experience of Bayer Corporation with electronic capture of field residue trial data. However, much of what is said about residue trial data applies equally to performance data. Major characteristics of field residue and performance trials are given. Advantages and disadvantages of electronic field trial data capture for study sponsors and field investigators are briefly discussed. The chapter concludes with some speculation on the future of electronic field trial data capture.

Field Residue Studies and Performance Trials

For those who may not be familiar with these types of studies, here are some major characteristics. Field residue studies are required by US EPA for registration of pesticides. The purpose of the studies is to set the limits of use, that is, define the maximum use pattern: the maximum rate, number of applications, interval between applications and PHI (Pre-Harvest Interval). Three to 20 trials are required per crop, depending upon the acreage grown and the importance of the crop commodities as human food. It is very common for registrants to use the crop group approach for registrations. Representative crops in a crop group are tested and the registrant gets a registration for all crops in the group. GLP and GALP (Good Automated Laboratory Practice) standards apply to these studies. Major registrants, such as Bayer Corporation, conduct as many residue trials as possible on their own research farms and contract the remainder to independent research companies which range in size from those that may have several research farms to husband and wife teams with only one site under their control.

EPA does not require submission of field performance trial data and GLP standards do not apply to these trials. Performance trials are conducted to determine the use pattern required for effective performance. Much effort goes into determining the *minimum* effective rate. Usually, about 30 trials are required per major pest. Much of Bayer's performance work is done on our own research farms or by our Field Development Representatives. Contractors are used for crops and pests we cannot handle internally.

The differences between residue and performance trials, as well as the GLP requirement for residue trials, make it difficult to design an electronic system that can handle both types well.

Advantages and Disadvantages of Electronic Data Capture for Study Sponsors

The advantages of electronic field data capture are prompting sponsors to adopt electronic systems. Electronic checking of field reports for completeness and the elimination of the need to check calculations yield major time savings. Electronic data summarization for inclusion in the final study report is also a major time saver. These major increases in efficiency not only save money, they may allow a submission for registration to be made sooner. Smaller, but significant, savings are achieved by electronic transmission of data, including study protocols, initial and final field reports, and status updates. Generation of field report files is speeded by cloning of the initial file and automated entry of sequential trial numbers.

Electronic systems are also very efficient for management of field trials. Bayer uses a spreadsheet to facilitate contracting tests, ordering test substances and payment of invoices.

There are also disadvantages to electronic systems. The cost of a commercial field trial data system is significant and multiple copies are needed. The cost for a sponsor to develop a unique system of its own would be even greater, and, the sponsor may not have qualified personnel. There are costs for system validation and testing, development and maintenance of standard operating procedures, and training. Possible additional costs include information system hardware and software and operating system upgrades.

Advantages and Disadvantages of Electronic Data Capture for Contract Researchers

Advantages for contract researchers are generally considered to be less than for sponsors, but there are significant advantages. Electronic checking for completion is instant. Linkage, entering a given bit of information only once with instant copying to every other location where that information is needed in the report, saves substantial entry and checking time. And, it eliminates a substantial source of frustration common with many paper reports. Electronic transmission of data from and to the sponsor is just as effective for the contractor as it is for the sponsor. Generation of progress report files for electronic transmission to the sponsor eliminates many phone calls and completing and faxing paper forms. Use of spreadsheets to manage trial work is convenient and effective. Electronic logs can be used to generate paper records for sponsors still using paper forms. Having a system required by a sponsor means a contractor can work for that sponsor. Some years, there may be a substantial amount of work.

The cost of an electronic field report system is the major disadvantage for contract researchers. A computer suitable for field use is required. For larger contract firms, multiple copies of the system, multiple field computers and a network may be necessary. Personnel must be trained. As with any new computer application, proficiency comes only with practice. Frustration of learning the application and its "quirks" replaces frustration of entering the same data in multiple places on paper forms. Not having a system means a contractor cannot get work from a sponsor that requires it

Bayer Experience with Electronic Capture of Field Data

We believe our experience can serve as an example of what is necessary to implement use of an electronic field data capture system. Bayer Corporation has been upgrading the quality our field residue studies since 1994. We implemented the new study guidelines in 1995; a year before required by the US EPA. We established a group to evaluate our methods and ways to improve them. The group determined that our paper field report system was the biggest problem we had. The group considered making major improvements in our paper report system versus an electronic system. The advantages of an electronic field report system made that the best theoretical solution. However, there was no commercial system available. We were starting to consider developing a system ourselves and possibly marketing it for use by other sponsors when Astrix Software Technology called about their new system called FieldNotes^{TM}.

We arranged a demonstration of the system and we liked what we saw. We were especially impressed that an experienced field residue study director and field investigator, who also had knowledge of computer system development, was the key member of the development team. Our experience developing our own electronic performance data capture systems had shown us the importance of someone with this dual experience.

After the demonstration, we consulted our toxicology group about development and validation of GLP data systems. These discussions confirmed our conclusion that we did not want to develop a system ourselves. Therefore, we arranged an in-depth demonstration of FieldNotes^{TM} and included our upper research management.

We did a vendor audit of Astrix. The audit team comprised our quality assurance supervisor, our information services representative and myself. We looked at system development, change and defect correction procedures and documentation. We also discussed our concern that Astrix was a relatively small company that could "go away tomorrow;" leaving us with what is sometimes called "vaporware." Astrix provided their standard lease agreement, which stipulates that in the event that Astrix cannot support the software in the manner specified in the agreement, Bayer would get the system code for the purpose of maintaining a useable system.

We then recommended the implementation of FieldNotes™ as our electronic field residue trial report system. Our recommendation report included a thorough cost/benefit analysis. Management approved implementation. Multiple copies of FieldNotes™ were leased; for our study directors, research farms, sample receipt and processing group, and field coordinator. We planned limited testing of FieldNotes™ against paper forms during the fall of 1996, but the pen computers we ordered arrived very late. Despite this setback, we decided to go ahead and implement use of FieldNotes™ on the seven Bayer research farms for our 1997 residue program. There were problems with the pen computers. They were too slow (486) and our farm personnel had problems with the effectiveness of the handwriting recognition program. Astrix provided excellent support and we got through the season without losing a single trial due to FieldNotes™, computers or handwriting recognition.

For the 1997 and 1998 residue programs, Bayer used study management companies because management removed responsibility for conducting and contracting field residue trials from our Field Development Representatives. It was thought that this additional responsibility would be too much for our one field coordinator (the author). All 1998 field residue trials were conducted with FieldNotes™. We encouraged contract researchers to acquire the system and many did. We were able to locate all trials with contractors using the system.

During 1998, EPA visited Bayer Research Park at Stilwell, Kansas. EPA audited one completed residue study and an application in a trial being conducted using FieldNotes™. There were no findings with respect to the use of FieldNotes™.

All 1999 residue trials were conducted with FieldNotes™. Additional contract researchers acquired the system. There is now a more than adequate number and distribution of contractors using the system to allow us to conduct trials on any crop grown in the United States and Canada. During 1999, Bayer implemented use of a spreadsheet to facilitate management of field residue trials. We discontinued use of study management companies for the 1999 residue program. Through efficiency gains due to use of electronic data capture and trial management systems, we were now doing almost twice as many trials as we had ever done before, and, we were doing them at a lower cost per trial.

Prior to and during 1994, we had been conducting 50 – 75 residue trials per year. During 1995 through 1998, we conducted 150 – 175 trials per year.

During 1999 we conducted approximately 300 trials. The Bayer residue program for 2000 is now in progress and we are conducting approximately 900 trials. This number of trials has required a great deal of extra effort from all Bayer employees involved with conducting and supporting the residue program. We've given that effort and we are succeeding, but this level of success would not be possible without use of our electronic data collection and trial management systems.

We are achieving the expected time savings in data checking and summarization. Last year, after one of our study directors had completed the final study report on his first residue study conducted entirely using FieldNotes™, he called me. This is an approximate quote: "Ken, you know I had my doubts about FieldNotes™, whether or not we would really save much time. Well, now I'm a believer. I just did in two hours what used to take two weeks."

Of course, there have been some problems. The primary problem has been missing data. We have always had problems with missing data in residue field reports. Some of these continued after implementation of our electronic systems. However, now FieldNotes™ has completion indicators on the table of contents form buttons. And, we have implemented a new policy for contractors. All electronic notebooks will be checked for completeness and will be complete before final payment is authorized.

Speculation on the Future

Even though we have solved many problems and made what I consider spectacular progress in just a few years, I believe we are just getting started with electronic field data collection. Further improvements will be coming and they will come fast. Improved importation of data from other electronic data systems has already started. FieldNotes™ imports weather data from a spreadsheet. Bayer is modifying its electronic performance data collection system to be able to import data from the commonly used commercial performance data systems. Data transmission over the Internet is improving and is already far superior to regular mail and even next day delivery services. There will be smaller, more rugged, and easier to use field computers. There will be headsets with tiny screens just in front of your eye so that the form image appears to float in space before you. The entire computer will be in a small headset. We will talk to the computer. It will seem to carry on an intelligent conversation about our experiments. Field scientists will feel like Captain Kirk. "Beam me up Scotty."

Chapter 7

Electronic Data Collection from the Study Director's Viewpoint

S. Scott Brady[1]

Aventis Crop Science, Research Triangle Park, NC 27708
[1]Current address: 5485 Southside Drive, Gulf Breeze, FL 32561

Presented below are the author's opinions on the utility of electronic data collection for use in pesticide residue field trials conducted to meet US EPA registration requirements. These opinions are based on over 30 years of work in the pesticide field.

In the years that I have been involved in industrial residue chemistry I have seen government requirements for data submission increase tremendously. Back in 1967, when I went to work for CIBA Agrochemical Company, one needed to establish only a few field trials, representative of the soil and weather conditions in the various parts of the country in which the product would be sold. Only a control sample and one treated sample from each treatment regimen had to be collected. Little data describing what was done in the field was required. At the laboratory, only one sample each of control, fortified control, and treated had to be analyzed; and, of course, Good Laboratory Practices had not come into vogue. Compare this to today's requirements set by EPA for up to 20 trials to be conducted for a residue-at-harvest study, with specific requirements regarding the regions in which the trials must be located, two samples to be taken from each treatment regimen, data to be collected on each operation performed at the trial site, weather data, chain of custody, storage and shipping conditions, data regarding handling of the samples at the laboratory from receipt to finished analysis, standards data, etc., etc., etc. Finally, the whole process must be conducted in compliance with GLP requirements.

Fortunately, we are now in the computer age. My first experience with computerized data came about when I worked for the Florida Department of Agriculture back in the sixties. Florida was one of the pioneer states regulating pesticide residues in food and has continued to be a leader. At the time I worked for the Department there were three laboratories in the Pesticide Residue Section, one in Tallahassee, one in Sanford (located in central Florida), and one which moved between Miami and Belle Glade, which is on the southeast side of Lake Okeechobee. Each lab covering about one-third of the state, analyzed vegetables, milk, animal fat, and stock feed produced in or shipped into the area. Our analytical reports were mailed daily to Tallahassee, where the results were transcribed into a mainframe residing database. Other reports were compiled from these data and printed for several uses, one of which was a hard copy that was sent to the growers for their records. One large vegetable and sugarcane grower, A. Duda and Sons, was ahead of its time in that they kept computer records on fertilizer and pesticide applications made to each acre of its land. Mind you, this was before the advent of the personal computer.

In 1970, I was hired by Sandoz Crop Protection to set up and run its Residue and Quality Control labs. A young woman chemist whom I hired later decided she would like to obtain a degree in computer science, a relatively new field at that time. She asked if she might take as a project for her classes the computerization of calculation and reporting our residue results. Of course I agreed. She wrote a handy little program that required submitting the analytical data on punch cards to the computer center at the University of Miami. One of us had to go to the computer center, punch the data onto the cards, submit the cards for input, and return either later that day or the next day to pick up the print-out. The entire procedure was probably more work than the manual method, but at least the print-outs were neat and it was a start on computerization. We later borrowed a remote terminal from the University and tried data input over the telephone line, but that setup did not always work. I was later transferred to the Sandoz national headquarters in New Jersey, where I was able to make use of the company mainframe for this calculation and printing.

My next experience with computerization of residue studies came many years later when the Astrix Software gentleman came to the AgrEvo Research Center and demonstrated their product, FieldNotes, to several of us. I was immediately impressed with the potential it had for making my job easier and my study reports better. Unfortunately, my supervisor at that time felt that there were GLP issues involved with electronic data collection and storage that could not be overcome. Luckily, he referred the Astrix folks to the world-wide AgrEvo functional head of Residues in Germany. Herr Dr. Uihlein became impressed with FieldNotes and set up a core team to study it and any available competitive products. I was appointed the North American representative to that team. At a meeting in Frankfurt the team chose FieldNotes as the standard software for worldwide use by AgrEvo, chiefly because there were no competitive products as far along in development. We in the U. S. have been using FieldNotes to an increasingly greater extent since that time.

My main complaint against contract Principal Field Investigators (PFIs) is their tendency to overlook the less important but still required details of the study protocol, failure to record all necessary data, and/or recording it in the wrong place. From my standpoint, the advantage of FieldNotes is that data recording is standardized and that for most screens, the operator is required to enter the called for data before he can exit the screen. Another advantage is that all data are placed in databases that can be queried in a flash in order to compile the tables that go into the final study report. Queries can be built to compile data in any combination one wishes. This saves an appreciable amount of time that would otherwise be spent searching through data books received from each trial in the study, in order to find the data that go into the report tables.

It is the stated objective of AgrEvo, and now Aventis, to bring one new product to market each year. This, of course, requires research on many more compounds than the one which reaches the market. Thus, a lot of data must be collected and reported. Most of our residue study final reports contain a lot of "boiler plate" type text. Macros can be developed which enter the variable text into new reports. Database queries, either in EXCEL or in Microsoft ACCESS, can be developed which almost instantaneously pull data into the tables that go into the report. And if the database is on a local area network to which the QA Unit has access, the QA review time should be much faster. These all make for much faster reporting time.

As many of you know, for some time now there has been talk of EPA eventually accepting electronic data submission for registrations. This, I have been told, is already being done in the European Union. Dr. Uihlein, whom I have already mentioned, sometime ago set up a database, called RESIDAT, for that purpose. He has directed that all North American residue data will also be entered into RESIDAT. The data we collect in FieldNotes can easily be transferred into RESIDAT.

There is now, to the best of my knowledge, one main alternative to FieldNotes for field data collection, namely American Ag's product

Advantage. I have seen Advantage demonstrated, but I must confess that I have not used it myself. However, I don't believe it is as far along in development as FieldNotes. I understand that ACDS Research and DuPont have also developed EXCEL spreadsheets that are used by the Principal Field Investigators for field data entry. In my view, the disadvantage of these spreadsheets is that data entry is not required before leaving a screen. Thus, some required data may not be entered by the PFI. Some people have voiced the opinion that it is good to have several choices in data collection software packages in order to gain the price advantages brought by competition. This maybe true, but on the other hand, it requires the field cooperator to purchase more than one software package, thereby increasing his cost of doing business. Of course, such cost increases have the greatest impact on the small field cooperator, several of whom do excellent work.

In conclusion, I foresee more and more computerized data collection and submission, possibly even submission to the sponsor over the Internet. If this can be made secure enough to satisfy our legal people and QA people, it would allow the Study Director much more real time monitoring of his studies. If the industry and the EPA will get together and standardize the data required to be collected and the format for the final report, life could be made much easier for the Principal Field Investigator, the Study Director, the sponsor QA auditor, and the EPA reviewer. This should in turn allow EPA to complete its submission reviews more expeditiously, and possibly allow new products to be brought to market faster.

Chapter 8

Auditing Electronic Field Data

Renée J. Daniel

Perspective Consulting, 1034 Greystone Lane, Sarasota, FL 34232

Historically all Good Laboratory Practice (GLP) compliant field data have been collected manually and documented in an organized manner in a Field Notebook prepared by the Sponsor or Field Management Company. The paper Field Notebooks have generally been well organized and logistically easy to audit. In the past several years software programs have been developed that are capable of generating GLP field data electronically. Programs currently in use are American Agricultural Services (AASI) Advantage™ and Astrix FieldNotes. Electronic data have provided a challenge for Quality Assurance (QA) Auditors who have been trained to audit paper data. They have also provided an opportunity for learning how to deal with the idiosyncrasies of auditing electronic data. There are several critical areas that need to be addressed when working with electronically generated field data.

Training

Management at a GLP facility must assure that personnel have adequate education, training and experience to operate the program and handle any unforeseen circumstances that arise due to the technology (i.e., software, hardware). Although training of GLP Study Personnel is extremely important, Management cannot overlook the fact that it is critical that QA Auditors receive training as well. Without thorough knowledge of software operation, QA cannot adequately audit the electronic field data. Attendance at an outside course (e.g., software developer) is highly recommended; however, training can also be obtained in-house from a previously trained individual if resources are an issue. During an audit, QA should assure that training is adequate and documented.

Equipment

Since GLP data must be entered directly and promptly, it is usually necessary to take a laptop to the field for data entry during events. Logistical issues must be considered such as durability, environmental conditions, adaptation for viewing the screen in sunlight, backup power and system security while in the field.

Laptop in Field vs. Desktop in Office

Although not as desirable, there may be situations where a desktop is used instead of a laptop such as in areas with intense heat or other environmental conditions, which preclude the use of a laptop in the field. In these cases, communication systems must be in place (e.g., radios, cell phones) to assure that data are recorded directly and promptly. It is more difficult to audit in these situations since QA has to assure that field procedures follow the protocol and Standard Operating Procedures (SOPs) and that the data are properly recorded. This may require QA to take notes in the field of the values recorded electronically which will need to be checked after the application/sampling. QA may also want to inspect more frequently such that the portion of the work conducted in the field is inspected separately from the data entry in the office. For the data entry inspection QA would listen to the communication of the values and assure that they are entered accurately and in the right area of the electronic notebook.

Power Considerations

Some laptop batteries may last long enough to perform calibration and application. However, there may be unforeseen circumstances with the application, which may require longer access to the laptop such as a re-mixing situation or waiting for the wind to diminish. Also, the battery may have been inadvertently not charged correctly which might shorten the available use time significantly. In order to assure that data are not lost, a source of backup power, such as an adapter that plugs into the vehicle cigarette lighter, should be available.

Backing Up and Transferring Data

Systems for backing up data are also required. There are a variety of systems, which will meet the requirement, and the decision should be based on logistics, resources and systems already in place. Some appropriate back up systems are floppy disks, zip drive, another hard drive, network server and CD-ROMs. Data should be backed up every day that original raw data are entered into the program. Ideally, backup data should be stored in a fire-resistant area or a location separate from where the primary data are stored.

Another equipment consideration relates to the transfer of data. Since study data will usually need to be transferred to the Study Director at regular intervals (typically after each application, sampling and shipment), the transfer rate is critical. Field sites, which are relatively isolated, may have poor telephone line quality and thus the upload can take hours. This will also be the case for downloading software updates. The upload/download time may be an important consideration if this will tie up the only telephone line or fax line. If available in the area, it is highly recommended that a faster form of transfer be obtained such as cable modem or DSL. Since telephone line quality is a common problem, it is recommended that the software developers send updates on CD-ROMs.

Hardware and Software Security

For hardware security, the laptop should be kept in as secure a location as a paper notebook would be kept, whether in the field or in the office. In addition, the environmental conditions are particularly important and temperature extremes and high humidity must be avoided.

Software security is maintained by using individual passwords. During a study inspection, QA should assure that the person entering the data is the person that has logged in to the system. If not, this is a GLP compliance issue since the person entering the data will not be the person identified as such in the electronic data. QA personnel should have "read only" access to the data in order to avoid any inadvertent changes.

Validation and User Verification

To conform to GLP requirements, all systems generating data electronically must be validated. For software currently in place, this is done in steps. The software developer thoroughly validates the program prior to its release. Sponsors also perform additional validation using their own protocols. The responsibility of the Field Site user is to verify that the program works on the system(s) to be utilized at their site. It is important to note that this verification must be done at the beginning of the season (prior to doing any field studies with the system) and following every software or significant hardware change. QA should check during an audit to assure that all systems have been adequately verified for the most current software build and that each verification has been properly documented. Current build numbers can be obtained for the software developer. In addition, QA should assure that the verification documentation is properly archived.

FieldNotes user verification can be done by following the SOP written by Astrix which can be downloaded from the Astrix website. The SOP contains a script that is typed into the computer system. If the printout matches the SOP script then verification is complete and a record of this should be placed in the facility archives. This process takes an hour or less. For FieldNotes, calculations are not verified at the field site since it is felt that the software developer and Sponsors have already adequately performed validation of these steps. The user is simply assuring that the program operates on their system(s). It is important to note that the SOP script must be typed in exactly as written in order to confirm the printout accuracy. This should be checked by QA.

For Advantage™, more extensive field site verification is conducted. A field site notebook is used which verifies every step in the data entry process. Following this procedure, a form is completed and returned to AASI where it is checked to assure that verification was properly conducted and documented. Due to the extensiveness of this verification, it takes approximately two hours to perform.

Maintenance and Repair Records

Although computer equipment is used differently than other field equipment, it is generating raw data so a maintenance logbook is required. All maintenance and repairs to the computer system should be recorded. For example, software and hardware maintenance records would include, but not be limited to, system/program updates, disk scans, and defragmentations (e.g., preventative maintenance). Repair records would include, but not be limited to, calls to the software developer about problems/bugs and steps necessary to fix any software or hardware problems. The documentation must include whether the procedure was routine and followed SOP and for non-routine must include nature of defect, how and when discovered and corrective action taken.

Standard Operating Procedures

Since there are procedural differences between handling paper and electronic data, additional SOPs are required. Some of the areas that should be covered are:

- Raw data definition;
- Assignment of users and access level;
- Security;
- Computer maintenance;
- Raw data backup method, frequency and storage location;
- Procedures to be followed in the event of computer failure or unavailability;
- Trial receipt and transfer of incremental and final data;
- Amendment/deviation documentation and notification procedures;
- Plot diagram generation;
- QA audit procedures.

Referring to a user's manual may cover some of the areas. Since some of the areas may overlap current facility SOPs, QA should assure that there are no discrepancies between electronic data SOPs and other facility SOPs.

The electronic data SOPs must have management approval, should be site-specific, where appropriate, and ideally should follow the format for facility SOPs. The electronic data SOPs do not necessarily need to be in the same binder or have the same numbering system as the rest of the facility SOPs. However, it is critical for GLP compliance that they be accessible to those who need to follow the SOPs and to auditors. QA should assure that all required electronic data SOPs are in place, approved by management and are being followed.

It is important to note that while in the past it was acceptable for contract field sites to use Bayer SOPs for FieldNotes studies, this practice will not work if FieldNotes is being used for another Sponsor. In addition, Bayer SOPs are lacking the SOP for site (user) verification, which needs to be in place at the Contract Field Site. If the Astrix SOPs for FieldNotes are downloaded, assure that all requirements of the SOPs will be followed. Ideally these SOPs should be used as a template with modification to meet facility needs. The downloaded SOPs should be carefully read prior to implementation to assure that there are no discrepancies between the SOPs and the facility SOPs already in place or the practices in place for studies with electronic data.

The AASI approach to electronic data SOPs is to provide guidance to the contract field sites, if requested. This includes direction on areas to be covered in SOPS and suggested content.

Auditing Issues

Preparation

In order to effectively audit studies where data are generated electronically, QA should create a special checklist or add electronic issues to a current checklist. The checklist should include items such as assuring that build is current and validated and that maintenance is documented.

Prior to conducting audits, QA should review the paper protocol plus any documented amendments and deviations. QA should assure that the paper protocol and amendments will be readily available during events in case they need to be referenced for information. The Field Investigator (FI) should enter amendments electronically if the system allows and this should be checked by QA. Any electronic entries previously made should be reviewed in advance if possible. It is a good idea for QA to print blank forms prior to going to the field since it may be hard to follow the screens during electronic data entry. For Advantage™, this can be done directly from the program and for FieldNotes blank forms can be downloaded from the Astrix website. Printing of blank forms in advance is also a good idea for the FI, in case there is a problem with the electronic system. QA should also assure that backups are being done as required by protocol and/or SOP and that any required incremental updates are being sent to the Study Director within the allowed timeframe.

Paper vs. Electronic Data Audit

Some QA Auditors prefer to audit printouts of the electronic data rather than the electronic data itself. This may be partially due to lack of familiarity with electronic auditing and partially due to unavailability of software to QA. There are several arguments in favor of QA auditing the electronic, rather than the paper data. One reason is because there may be perceived compliance issues related to the printouts that do not exist in the electronic data (e.g., initials do not print on hard copy). Another is that QA may not be given printouts of all of the pertinent data such as site logs. If QA does choose to audit the printouts, the site logs, audit trail and notes pages must be printed and audited as well. A third reason is that the FI does not usually organize printouts as well as a paper notebook would be organized which may cause confusion. In some cases, the printouts may not indicate the actual plot identification (e.g., treatment 3), which could lead to confusion when auditing a study with several plots treated the same day with different rates.

In order to audit the electronic data, QA must either have their own copy of the software or use the computer where the software is installed. In the case of FieldNotes, the latter may be more practical since the former would require the purchase of the program for QA. With respect to Advantage™, the program is available without charge to any Quality Assurance Auditor that needs it so resources should not be an issue. When auditing the electronic data, it may be somewhat difficult to follow the whole study since documentation cannot be spread out on a desk. For this reason, it is recommended that certain pages be printed out for easier reference such as the audit trail and notes pages.

Critical Areas to Consider

The software programs perform calibration/application calculations and the calculation process is validated/verified. However, for added comfort level with the programs and the data, it is strongly suggested that QA conduct an independent check of calculations.

There are some electronic data that are directly entered by the FI and some that are entered automatically by the program (e.g., chronological log entries). QA should check not only the directly entered data, but also the computer-entered data. In order to conduct a thorough data audit QA should make an effort to determine which data for each software program are direct entry and which are automatic.

During a raw data audit, QA should assure that notes and various descriptions (e.g., sampling method) are clear and thorough. In addition, the audit trail should be checked for clarity and to assure that the original entry can be determined as required by GLP. Also, it is critical for QA to assure that all entries "make sense", especially in connection with event times and dates.

With electronic notebook studies, there will be paper data to audit as well. This will include facility data (e.g., weather data, equipment maintenance records, temperature storage logs) and study specific data (e.g., faxes, emails, paper notes). If any of the paper data has been transcribed into the electronic notebook, this should be checked by QA to assure accuracy. QA should assure that all paper data have adequate identification (e.g., study and trial numbers) and that all of the pertinent paper data or exact copies are sent to the Study Director for archiving.

Potential Concerns

Plot Maps

Many plot maps are generated in a separate program (e.g., Microsoft Paint) and imported into the field software program. If there is any information on the electronic plot map that was not written elsewhere first (e.g., plot dimensions, slope, row direction), the electronic plot map becomes original raw data and entries must not be obscured if changes need to be made.

Current software programs do not allow for easy viewing of a previous plot map once it has been changed. In FieldNotes, for example, the audit trail will indicate that the change is from an old image to a new image. The software developers may be able to retrieve the old image (plot map) in the program; however, it is not readily accessible to the FI.

For Advantage™, plot diagrams must be created in a separate program and imported. If changes need to be made, the diagram must be removed and re-imported which will create an audit trail indicating that the diagram has been replaced. The file for the previously imported map must be retained in order to view the original entry.

If a slope change is made on the plot map in FieldNotes, the audit trail will have a numerical entry for the change (e.g., from 90 to 200 instead of E to W). This is due to the fact that the plot map is a picture and not text. The software developers will know how to translate the numbers but this information will need to be available during EPA audits as well.

There are ways of which the FI and QA should be aware to make the plot map creation and modification procedures GLP compliant. Any original data (e.g., dimensions, slope, row direction, distance to markers, wind direction, buffer distances) should be entered first in text form into the field software where an audit trail will exist. If this is not possible (i.e., program design does not allow it) the original entry can be made onto a properly labeled sheet of paper or form that will be sent with the raw data. If this does not occur and actual raw data are entered first into a program, such as Microsoft Paint, this fact should be included as an exception on the compliance statement.

For GLP compliance it must be assured that when the plot map needs to be changed, the original entry can still be determined during an EPA audit. This can be done by printing the original plot map prior to changing it and sending the printout (with explanation) along with the raw data. If this is not practical, a note should be attached to the changed electronic plot map indicating the exact original entry (e.g., row direction changed from N-S to S-N).

If a map is scanned and imported into the program, the original scanned map is the raw data and should be saved either as facility data or study-specific data, as applicable. When a plot map is drawn separately and imported, QA should assure that the imported map matches the original (e.g., all lines visible).

In situations, where some of the imported map is not clear, the original electronic (e.g., GIF) file should be printed and saved in case of a future EPA audit.

Data Entered Late

If there are any situations where data need to be entered late and will appear unusual, these instances need to be thoroughly explained. For example, the battery dies on the laptop, there is no adapter available and only part of the calibration was done before the laptop shuts off. The FI will be transcribing some data later that day or night. The program will time-stamp some data as the event occurred and some data at the time of transcription. The differences in the entry times will be atypical and should be explained to prevent perception of "creating data in the office". The explanation can be done simply by attaching a note to the transcribed data indicating the situation and referencing the source of the transcribed data.

Test Substance Tracking

For FieldNotes, test substance tracking may be difficult when one container is used for several studies. The general log for test substance will automatically track usage at each application. For this reason, if QA is auditing the individual study data the numbers will not seem accurate. For example, 50 grams of chemical A is received. Study 15 application is made on 07/08/00 and 10 grams of test substance are used. The general log will indicate that 40 grams remains. Study 25 application is made on 07/10/00 using 20 grams so this amount is subtracted from the chemical inventory in the general log. When the FI goes to make the second application to Study 15 on 07/15/00 the amount of chemical remaining will be indicated in the log as 20 grams before the application is made. Since only 10 grams out of 50 were used for that study, QA may expect to see 40 grams remaining instead of 20 grams. If QA is trying to track use, this may be confusing. QA should be made aware of this during the audit. Consequently, it is highly recommended that a separate container of test substance be sent for each trial and/or study in order to facilitate tracking during the audit.

Generic Menu Choices

Many of the entries in both FieldNotes and Advantage™ are made using drop-down menus. The choices are not always the most accurate for the situation at the field site (e.g., sprayer types, nozzles, crop stage). QA should check during the raw data audit to assure that the choices made are an accurate reflection of the practices utilized. It may be necessary to attach a note to the pertinent page to clearly explain certain situations such as unique nozzles or different sized nozzles for airblast applications. Explanations may also be required for airblast applications. Due to the variety of airblast calibration procedures utilized by field sites, the electronic forms may not correspond with the way calibrations are conducted at the site For example, if the form only allows for entry of total calibration output volume but the FI collects individual nozzle outputs, this should be indicated in an attached note including documentation of the individual nozzle outputs. This situation may also occur if an in-line flow meter is used for airblast calibrations where the initial volume is not typically noted. If the form requires entry of initial volume, this entry can be given as an estimate with a note attached explaining the situation.

Conclusion

QA must be aware of the issues and requirements for electronic data in order to adequately assess data accuracy and GLP compliance during inspections and raw data audits. Several significant questions to keep in mind are "What are the original raw data?", "Are the electronic data complete, accurate, and GLP compliant?" and "Have all of the original raw data or exact copies been sent along with the study data for archiving?"

In conclusion, as technology advances and the use of electronic notebooks for field studies becomes more prevalent, QA Auditors will need to comprehend not only the technology but also specialized electronic data auditing techniques: Techniques that may be critical to the GLP compliance of the study.

Chapter 9

Future of Electronic Good Laboratory Practice Study Management Is Now

Fate Thompson, Tommy Willard, and Carla Wells

American Agricultural Services, Inc., 404 East Chatham Street, Cary, NC 27511

Advantage™ Project Management Assistance, including Advantage™ Field Trial Manager and electronic Field Trial Notebook, is a complete electronic GLP study management system. It utilizes an Access database and Visual Basic and has resources for test substances, formulations, test systems, and study management, laboratory, and field personnel. Advantage™ Project Management Assistance automates the creation of studies, trials, sample numbers and labels, and electronic field trial notebooks. The Field Trial Manager module is on CD for installation at field test facilities. Electronic Field Trial Notebooks are generated for transmission by e-mail or diskette. Trial data is received from the field by e-mail or diskette. The Project Management Assistance module automatically creates protocols, status reports and study reports in your format and word processor of choice (Word, WordPerfect, etc.). Advantage™ Project Management Assistance, Field Trial Manager and electronic Field Trial Notebook are fully validated and fully compliant with GALP and GLP regulations.

Introduction

In reviewing the preliminary Table of Contents for this ACS volume, it was noted that there were many very good topics on capturing field and laboratory data, reporting of results, and electronic submission of data. Each of these is an important component in the overall process. However, no broader view of why the agrochemical industry should consider taking an electronic approach, nor where we are going in the future was to be presented. Therefore, I talked with the editors and they graciously permitted me to write this chapter entitled *The Future of Electronic GLP Study Management is Now*. I ask your indulgence as I will periodically site Advantage™ Project Management Assistance software as a point of reference since it typifies key aspects of why an electronic approach is important in this industry and where we as an industry are going.

As we look across the industry today, there are at least three systems involved in electronic capture of field data including Advantage™ eFTN, FieldNotes, and various Excel Spreadsheets. Among private companies there are a number of systems available for electronic capture of analytical data and systems for compiling submissions to allow for electronic submission to regulatory authorities. In addition, across the industry there are various configurations for residue databases in operation among sponsor companies. However, there is only one fully-integrated, electronic GLP study management software system in operation today. That is Advantage™ Project Management Assistance.

The heart and soul of the process is not data capture or a database, but a fully-integrated management system built specifically for the study director or project manager. The study director is critical to this management process. Advantage™ has been specifically designed to aid him. As noted in Table I, we ask a tremendous amount of the study director.

Table I. Potential Study Director Management Functions

▪ Bid Requests	▪ Contractor Payables
▪ Protocol Development	▪ Sample Numbers
▪ Sample Labels	▪ Field Trial Notebook
▪ Monitor Progress	▪ Management Reports
▪ Receive Field / Lab Data	▪ Final Study Report
▪ Query Database	▪ Electronic Submission

Background

Since 1987, American Agricultural Services and its Lyon, France subsidiary, European Agricultural Services have managed approximately 700 GLP studies and 8,000 GLP trials worldwide. Most of these were managed while serving as the study director. Since 1990, computer programming has been actively underway to develop a management system designed specifically to aid the study director. The driving premise has been to allow one time entry of data into a fully integrated system which operates on a database and facilitates study management thus reducing time and cost. The key development considerations that should factor into the creation of any electronic study management tool are economics, overall functionality, quality, and consideration for future requirements. Before elaborating on each of these in further detail, it is beneficial that we provide a brief overview of the functionality and capability that should reside in project management assistance software utilizing Advantage™ as an example.

The Advantage™ software suite consists of two basic components. The first is Project Management Assistance. It is utilized by the study director and resides at the agrochemical company or study management company. The second component is the Field Trial Manager and electronic Field Trial Notebook that are utilized by the field research contractor and resides at his office and out in the field. A later chapter entitled Development of Advantage™ eFTN: A GLP Field Data Capture System will specifically address the functionality and capability that resides in Advantage™ Field Trial Manager and Advantage™ eFTN. The balance of this chapter will specifically address Advantage™ Project Management Assistance as a study management system which, as indicated earlier, is specifically designed for use by the study director and is the heart and soul of the study management process.

Advantage™ Project Management Assistance

Advantage™ PMA is designed to function on a network such as Windows NT that allows multiple users access to the software at the same time. Alternatively, it can be installed on and used on an individual PC. The major functional components and the role that they play in the fully integrated process are outlined in Figure 1 and the chapter entitled "Development of Advantage™ eFTN: A GLP Field Data Capture System."

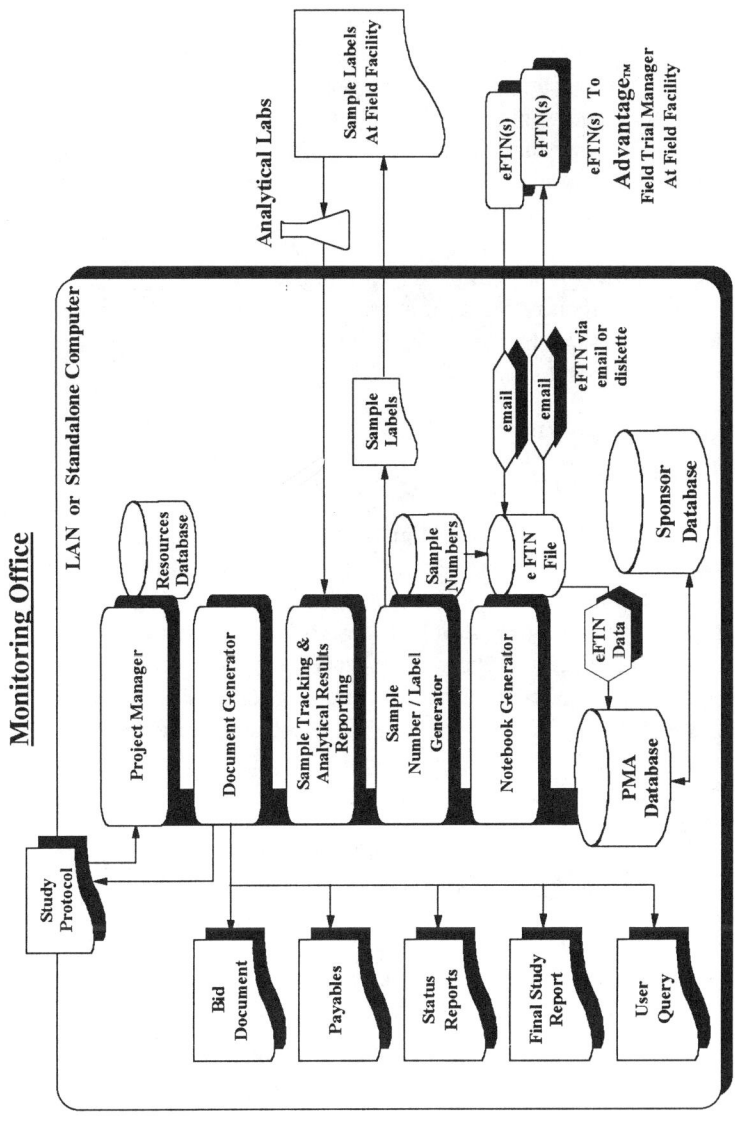

Figure 1. Advantage™ Project Management Assistance Software Configuration

Project Manager

Initial input to establish basic study design is made into this module. This includes selection or entry of study type, test substance, specific formulations, number of trials, trial locations, and determination of key study people such as the study director, field contractors, and supporting laboratories. Throughout Advantage™ PMA, a resource database is utilized which has previously been loaded with background information that can be accessed simply by point and click and does not require repeated entry. This includes, for example, name, title, address, phone, fax, e-mail, etc., for study director and key field research contractor or laboratory personnel. By entering the study type, Advantage™ recognizes the basic requirements for a crop residue versus a soil dissipation or groundwater or exposure study and automatically sets up files in the database to anticipate and to receive the type of data to be generated.

Protocol Generator

Once certain relevant information has been entered into the Project Manger module (as well as the Sample Number Label Generator and Field Trial Notebook Generator), the study protocol can be generated electronically. By establishing a protocol format of your design and utilizing the functionality within the protocol generator, the study protocol with complete text and tables can be generated in one to two minutes simply at the push of a button. The protocol is available to you in your own word processing software and both the format as well as the finished document can be created and edited by the user without additional programming input from American Ag.

Sample Number Label Generator

As this is a fully integrated system, certain input that has previously been made into the Project Manager module is automatically received into the Sample Number Label Generator. The user is then able to define, for example, number of treatments, plots, sub-plots, number of replications, and determine specific sampling events, i.e., as for a harvest study versus a decline study. This module then generates sample numbers for the entire study, trial by trial, which can then be exported into a LIMS system. Alternatively, sample numbers can be imported from a LIMS system rather than being generated within Advantage™. A sample list is generated which can be printed at anytime. In addition, sample labels are generated which can be printed with all of the information needed to clearly describe that sample with the single exception of the sampling date which is not known until the actual sampling event occurs in the field. These sample labels can be generated to contain any one of several bar code systems. These sample numbers also flow automatically into the Field Trial Notebook Generator.

Field Trial Notebook Generator

As input is made logically and sequentially into each of the previous modules, relevant information electronically flows into the Field Trial Notebook Generator in preparation for creation of specific field trial notebooks. Within the Field Trial Notebook Generator, another layer of detailed information is entered including, for example, application details of rate, timing, number of applications, method of application, etc. Advantage™ automatically creates or builds the electronic Field Trial Notebook by selecting the specific forms or pages which will be required for a crop residue, soil dissipation, groundwater, exposure study, etc., based on the study type that was indicated previously in the Project Manager module. All information that is known about that specific trial is automatically transferred from Advantage™ into each field trial notebook. The electronic Field Trial Notebook can then be sent by e-mail or diskette to the Advantage™ Field Trial Manager module that resides at the field research contractor's location. Advantage™ also automatically prepares the database to receive field biology data when it is e-mailed from the field research contractor back to the study director.

Sample Tracking and Analytical Results Reporting

The STARR module resides at the analytical laboratory whether it is internal to the sponsor company or is a contract laboratory. When samples are received at the analytical laboratory, barcode readers are used to acknowledge receipt of the sample and STARR automatically registers all information contained on the sample label to include the date of receipt of the sample. In addition, condition of the sample can easily be entered into the system and sample location can be easily specified. By using a barcode reader system as the sample moves from preparation to extraction to analysis to reporting, STARR automatically tracks sample movement throughout the laboratory facility. Having identified and tracked the sample electronically, STARR facilitates the summarization of analytical data into a final study report.

Project Management Assistance Database

The database that is resident within Advantage™, or the sponsor company database to which Advantage™ is linked, is a key to the overall functionality of this fully electronic, fully integrated system. Information from each of the previous modules and from the data collection in the electronic Field Trial Notebook or STARR in the analytical laboratory is continually accruing into the database. Typically, Advantage™ is interfaced directly to the sponsor's database. Alternatively, it can operate with its separate internal database. Based on the functionality of the Document Generator that is inherent in Advantage™ and the linkage of Advantage™ with the sponsor company's database, any number of reports of the sponsor company's design can be readily generated. Typical examples are outlined below.

Final Study Report

Recognizing that each sponsor company has unique study report formats and recognizing that the sponsor companies want the flexibility to internally modify their report formats, Advantage™ has been designed such that reports are generated in the report template or format designed and preferred by each individual sponsor company. This includes development of both text and tables as well as creation of the report in the word processing software preferred by the sponsor company. Since it is generated in a word processing format, the study director can easily edit the document. This process is fully automated such that having selected the study of choice and a previously designed report template, pressing a button allows drafting of the complete study report in a couple of minutes.

Bid Document

The same Document Generator functionality that was utilized to create protocols and final study reports can be utilized to create bid documents to secure quotations from subcontractors. As indicated previously, the format is of the sponsor company's own design including tables with key technical parameters and any supporting text. Relevant information that has been previously entered into Advantage™ when creating the study design can readily be pulled from the database to create the bid document. Once the bid document is created specific to that study or specific to that individual field trial, it can be sent by e-mail to the subcontractor. After the subcontractor enters cost and other relevant information into the document, it can be returned by e-mail back to the study director and the information integrated into Advantage™.

Payables

Sponsor companies have clearly indicated the desire to have the ability to monitor payables for contracted or in-house projects without being dependent upon in-house accounting systems. The Payables Module within Advantage™ accomplishes this by tracking by study and by trial the contract amount, when payments are due, specific amount of each payment, the total amount invoiced, the amount paid, the amount remaining to be paid, etc. Based on sponsor company preference, this can be run independently of company accounting systems and / or can be linked directly to sponsor company accounting systems.

Development Considerations

Economics

A key consideration in the development of Advantage™ software was to maximize study management efficiency resulting in reduced time requirements and significant cost savings. It was noted previously that utilizing individual PCs and paper field trial notebooks required considerable repetitive entry of the same information. The end result was that 80% of the study director's time was devoted to largely clerical activities. This was because development of the protocol, field trial notebook, status reports, study reports, etc., were stand alone activities that were not integrated.

As noted previously, the operating premise behind Advantage™ was a single entry of information into a fully integrated system such that all components are linked and any information entered into one component automatically flows into downstream components. This immediately refocuses the utilization of the study director's time from largely clerical to largely technical management and decision making functions -- which is the intent for this critical position. From American Ag's in-house experience, this transition from largely paper systems to fully integrated electronics systems with Advantage™ have resulted in up to an 80% increase in efficiency in study management time.

As a bonus, and as part of the intentional design, the complete study data now resides in a readily-accessible electronic database rather than sitting in a file cabinet on paper. This improved manpower utilization allows resources to be refocused toward improving the science, responding to internal and regulatory data requests, reducing overtime and submitting registrations on time allowing quicker entry into the market.

Functionality

The basic functionality consideration driving the development of Advantage™ was to mechanize and fully integrate all study management functions. As noted earlier, input of data one time was a key consideration. This was accomplished by developing resource databases that reside in Advantage™ Project Management Assistance and allow a myriad of information to be pulled into any given study simply by pointing and clicking on the relevant resource database. This includes items such as a list of all possible study directors and their address, phone, fax, e-mail, etc., as well as complete lists of all subcontractors and their contact information. Also included are lists of all sponsor company active ingredients and including all the supporting technical parameters for each. Having entered this information into Advantage™ one time, it is available without any retyping by simply pointing and clicking on this information. Once it is pulled into the specific study of interest, it automatically populates the data fields where that particular piece of information should go from development of the protocol through management of the study into development of the final study report.

User-friendly screens incorporated into Advantage™ give it a very familiar look. The design and functionality have a Windows look and feel. A tab layout is utilized to allow easy selection of the module within which one wishes to work, as well as the functionality within a module. This tab layout has the feel of chapters within a book, allowing the user to know exactly where he is and where he needs to look for a desired module or functionality. Movement backward or forward is a simple keystroke command. Pull down menus are integrated throughout the software that allow the user to tap directly into the resource databases to incorporate information by point and click rather than retyping. For example, all of the RAC test systems as defined by EPA and OECD, such as corn, citrus, coffee, etc., are readily available in pull down menus. Selecting the test system from the pull down menu also automatically identifies the standard sample fractions and places those in the appropriate location throughout the Advantage™ software. Button commands are incorporated throughout the software to allow use of a single keystroke to select pre-programmed capabilities such as switching from English units to metric units, cloning a study, generating a final study report, etc.

As the user navigates Advantage™, it is clearly apparent that it has been designed for an understandable yet flexible workflow. By simple point and click, the user can readily maneuver back and forth among the modules that have been previously described without being locked into a rigid sequence of events. Drill down capability allows the user to progress logically from project into individual study within a much larger project and into specific trials within a study. Within a trial, this same logical drill down sequence allows progression from general information into detailed treatment information and further into very detailed sample information, if desired.

Quality

GLP and GALP compliance as well as proper software validation are two of the factors that define quality in any computer software. Key compliance features incorporated into Advantage™ include requirement for a user name and assignment and entry of a secure password in order to access the Advantage™ software system. In addition, security levels can be assigned by the systems administrator that define the access level of the user as administrator, enter / edit, view only, and no access. In addition, these security levels can be assigned for each specific software module inherent in the Advantage™ system. Once any information is entered into Advantage™ and is saved, any change of this information requires creation of a Data Revision Record that indicates the field of information which is to be changed, the previous information, the new information, and the reason for this change. Also, automatically recorded are the user's name and the date and time at which this change occurred, as well as entry of the user's secure password.

Proper software validation is an integral part of any software development project. Validation at both a developer level and a user level will be discussed in much more depth in later chapters. In developing Advantage™, American Ag followed standard industry validation and documentation practices for development SOPs, a thorough and well-documented software development life cycle, authorization for commencement and completion of all development milestones, user requirements, system design specifications, functional design specifications, system integration testing, and validation testing. Since American Ag is a GLP study management firm, we have also validated Advantage™ in its entirety within our study management facility as a user. We leave user system validation by sponsor companies to their discretion, but are available to support and assist them as they may require. In addition, American Ag has and will continue to fully support vendor audits as required and requested by sponsor companies who license Advantage™.

Future Requirements

From its inception, Advantage™ Project Management Assistance software has been designed to allow for increased functionality and interfaces as they are developed in the future. These considerations for future requirements should be part of any decision on GLP study management tools. These include the development of a modular system that allows use of all or only certain parts of Advantage™ and the integration of these functional components with existing in-house capability.

It has been designed on a database platform for increased functionality allowing ease of drill down to more and more detailed information, as appropriate. This also facilitates the ability to query data from the database and to interact with other databases. Development is currently underway on a Document Generator module within Advantage™. This incorporates the flexible report template design and the speed and ease of report writing noted previously with the ability to easily mine data from virtually any database. The web basing of Advantage™, noted below, renders its operation database independent allowing complete compatibility with Oracle, SQLServer, etc., databases.

Current operation is on network operating systems such as Windows NT, which is by design a universal system. Building Advantage™ as a network-enabled system allows the user to install it on and integrate it directly into an existing network.

Last, but certainly not least, from its inception Advantage™ has been designed and structured with web basing in mind. We believe that this technology holds great promise for the agrochemical industry. This enables users access to Advantage™ any time from any computer anywhere in the world. Agrochemical companies can then globalize their operations by linking all their facilities and sub-contractors on a single web-based system. By early 2001, Advantage™ Project Management Assistance will be web based. Development of the web based Advantage™ eFTN will occur later in 2001.

Summary

Electronic study management in a fully integrated system, such as Advantage™, affords major time and cost savings. Whereas electronic data capture at the field and laboratory level are important, they are of limited consequence unless the efficiency gained from full integration can be established from data capture to final reporting and, potentially, electronic submission. This in turn translates directly into earlier regulatory submissions, faster response time to regulator's questions, earlier registration approvals, and quicker entry into the market. Electronic study management is the nexus of data inputs and outputs. Therefore, the maximum benefit to the study director, and in turn to the agrochemical industry, is obtained when it is a considered dimension of your decisions as they relate to the complete gamut of field data capture, summary analytical data capture, residue database output, and regulatory submission requirements.

Chapter 10

Automated Data Collection in the Laboratory Using the Chromeleon™ System

William H. Harned

Uniroyal Chemical Company, Crompton Corporation, 199 Benson Road, Middlebury, CT 06749

Automated electronic data capture systems have become increasingly important in the laboratory for ease of manipulation and reporting of chromatography data and also for generating a permanent audit trail of the parameters employed in the collection and analysis of that data. An audit trail is critical for regulatory compliance. The Chromeleon™ system from Dionex is a powerful tool for collecting data and controlling a wide variety of chromatographic equipment and detectors. It is operable from a laboratory PC, either directly or through a LAN or WAN system. A system administrator can control a hierarchy of privileges for access to the raw data. Prudent delegation of privileges allows a wide variety of personnel to view data, use methods or generate reports while restricting the ability to edit methods to the study director. The audit trail capabilities maintain a chronological record, with electronic signatures, of all people accessing or editing files.

As electronic data capture systems have become prominent in the research and analytical laboratory, they have allowed rapid and efficient acquisition, manipulation and reporting of vast amounts of scientific data. In addition, they have provided a means to generate a permanent audit trail describing the conditions under which a chromatogram was obtained, documenting the analyst performing the work and recording the precise time and date that the chromatogram was produced. Even the time of occurrence for each event taking place during the chromatography run, for example, a gradient or temperature change, can be documented precisely.

The Dionex Chromeleon™ data capture system allows the user, with appropriate drivers, to set the parameters of a chromatographic system, operate that system, collect the data generated by the system, store the data securely and manage the presentation of the data. Data collection is typically accomplished by converting the analog output from the chromatograph to digital data readable by a desktop computer. On older systems, this was accomplished primarily by use of a data conversion box. More recently, analog to digital or A/D cards have been used to convert the analog signal into a digital form. The digitized data are then sent to the PC for processing and storage. Figure 1 shows a typical HPLC system with all components controlled via the data system. Some newer instruments are capable of being controlled by a PC with their data output sent back to the PC directly through use of a serial interface card. Information stored on the hard drive of the PC can then be sent to a LAN server where it may be accessed by other authorized users. In most institutions data on the LAN server (Figure 2) are backed up onto tape or optical media on a regular schedule. The backups should be maintained at an offsite location, adding another level of security for the data.

Choosing a Data System

The decision on whether or not to purchase an automated data collection system for the laboratory is no longer based simply on cost. Once it could be considered something of a luxury, but in today's analytical laboratory it has become a necessity. With the advent of high throughput analyses, the speed and efficiency of data handling, and the ability to run chromatography and collect and process the data overnight on an unattended system, few laboratories can compete successfully without the advantages of automated data capture.

Perhaps an even more compelling advantage of the automated data collection system has appeared with the advent of the Good Laboratory Practices (GLP) standards (1), and the more specific Good Automated Laboratory Practices (GALP) (2). The automated data system now allows an electronic record of not only the chromatogram (output) generated during a chromatography run but also a record of the inputs (date, user, sample name

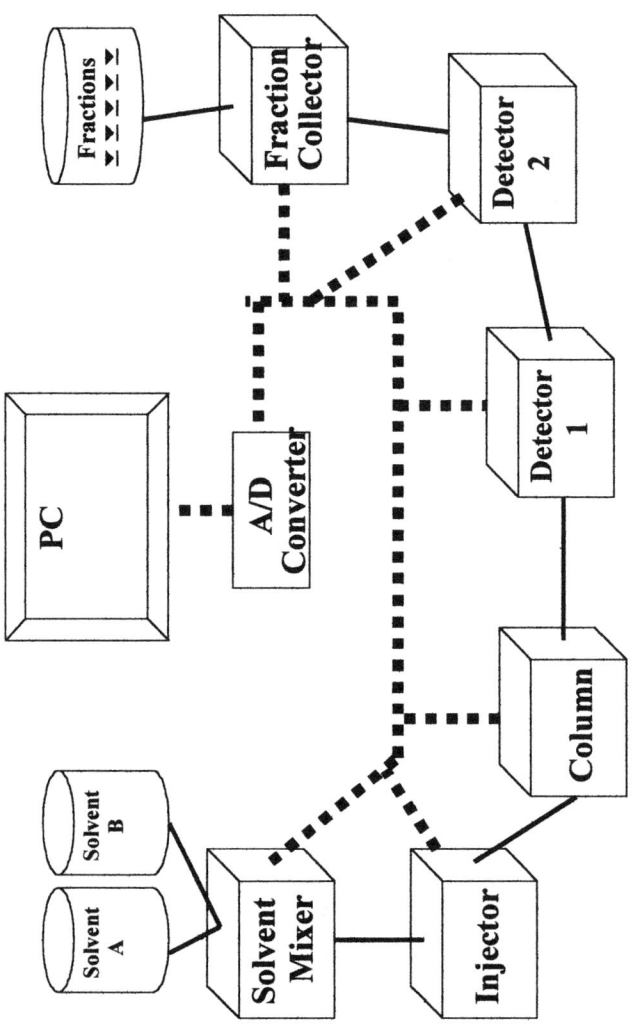

Figure 1. HPLC system control from a workstation.

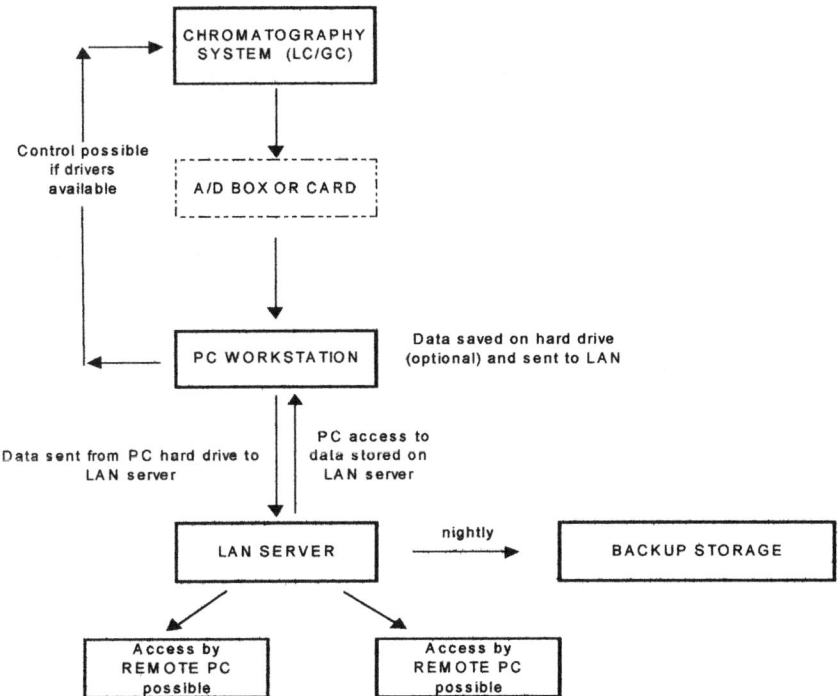

Figure 2. A typical Hardware configuration for data collection and storage.

and number, instrument, column type, temperature(s), flow rates, etc.). A dynamic audit trail can be kept for all settings, changes and results. When a change is made in an input parameter, a description of the change(s) including the original data, time of the change(s) and person who made the change(s) are recorded electronically. It is also possible to manually enter additional items or comments such as column pressure at any given time. On a properly administered system, this record cannot be altered and thus becomes a permanent record of the entire chromatography run.

Many considerations should be made when purchasing an automated chromatography data collection system. One must be very clear on what needs are to be met. It can be safely said that no single system is the best for all situations. However, the search can be reduced by prioritizing needs and requirements. Some of the major items to consider are :

- Can the system provide data in the form that is required for the tasks at hand?
- Will it work with the equipment on hand - chromatographs, PCs, etc.?
- Is there flexibility and expandability to meet anticipated future needs?
- Can the system be used as a stand-alone unit as well as a networked one?
- Is the system user friendly or will there be a large learning curve?
- Will the system allow me to meet GLP/GALP requirements?
- What kind of reputation does the manufacturer have - reliability, service, etc?
- Can I afford it?

Each of the above items must be investigated prior to making a purchase. One of the primary reasons for selecting the Chromeleon system was its ability to control most of the instrumentation already present in our laboratory which had been manufactured by a variety of different firms. Of course, the ideal situation is that the data capture system and chromatography systems are produced by the same manufacturer. Then one has a complete system comprised of components specifically designed to work together. In most cases, however, a substantial amount of equipment from other manufacturers has already been purchased and utilized in the laboratory, and budget conditions will preclude a complete replacement of working chromatography units.

The server requirement must also be a consideration. How many separate timebases (chromatography systems) can be controlled from a single server? A separate PC for each timebase avoids difficulties, but often many timebases must be served or controlled by a single PC server. In this case the ability to serve and/or control up to four, six or as many as 16 units becomes important.

System Qualification

In a GLP compliant laboratory, a data system must also meet explicit requirements guaranteeing the validity, quality and security of the collected data. Chromeleon has specific programs and templates for installation qualification (IQ), operational qualification (OQ) and performance qualification (PQ). IQ is normally performed by the vendor during installation of the system hardware/software onto the PC. The Chromeleon system has resident IQ Manager software that automatically checks the current installation after each new software update. A printed report can be generated showing system information, status of Chromeleon and all copied files, which are compared to a write-protected listing maintained within the Setup program. If there is any difference, a warning is issued.

OQ must performed after any new devices are installed in the system and whenever service or repair is carried out. The role of OQ is to demonstrate that the instrument functions according to the operational specifications in its current laboratory environment. If the environmental conditions are highly variable, OQ should be checked at the extremes as well as at normal ambient conditions. Chromeleon has a built-in OQ Wizard which the user can run whenever an OQ is deemed necessary.

PQ must be performed following any new installation and whenever the configuration of the system has been changed. PQ demonstrates that the instrument performs according to the specifications appropriate for its routine use. Chromeleon generates a set of templates to perform the PQ. The user enters the system specifications. For example, on the PQ sheet "Inj. Prec. and Ret. Repro.", the user enters the specifications that are required. The system checks the injector precision and retention time reproducibility, and if the result is within the specified limits, the report indicates 'OK'. Similar checks are available for injector carryover and linearity, pump gradient accuracy and reproducibility, detector noise and drift, detector wavelength accuracy, and detector linearity. A number of additional parameters may also be checked using the PQ templates.

Access Control

One of the most powerful features of the Chromeleon system and a necessary feature of any automated data collection system is a means of controlling access to the data during its generation and after its storage. Chromeleon accomplishes both tasks through privileges granted by its system administrator. The network capabilities allow a user to perform any allowed operation from any workstation (client) on the system. The data management system is a control level in addition to the overall network management system already in place on the entire local or wide area network.

The Access Control feature guarantees that only authorized users may gain access to the specific workstations, timebases, raw data, datasources, folders, run sequences, reports, etc. The operations that are access-protected include:

- creation, modification and deletion of users, datasources (databases), folders (directories), sequences, samples, worklists, workspaces, queries, server configurations, timebases (individual chromatography systems), libraries, control panels, report definitions and quantification methods
- saving and/or deletion of peak manipulations and results
- starting batch acquisition or batch processing, exporting and printing batch data
- backup and restoration of datasources, folders and sequences
- import and export of data
- access to servers and timebases

Under a program of access control, an administrator, who is chosen by management, grants privileges to authorized users, thereby defining the scope of the functions that are available to each user. The administrator also assigns individual passwords to each user. Whenever an individual user logs onto a workstation that is subject to access control using his or her password, all of the allowed "personal' privileges become available.

Chromeleon provides two separate password-protected programs to the administrator for control of access. The first program, CMUSER, permits the administration of User Privilege Groups and Access Groups. The possible operations are:

- Creating and editing access groups (A-Groups)
- Creating and editing privilege groups (P-Groups)
- Creating and editing new users (Chromeleon Users)

The second program, CMSECURE, serves to create a database that contains the status and all of the granted rights for each user.

Users can work with an item if they are members of an A-group that has been granted access to that item. The actions that they are allowed to perform on the item are determined by the assigned privileges (privileges assigned to members of a P-group).

As a very elementary example, suppose that a particular system has two attached timebases, T1 and T2. There are five users, U1, U2, U3, U4, U5, for timebases T1 an T2. User U1 is the system administrator and has unlimited access to all objects on both timebases. Users U2 and U3 are granted authorization to generate and record data using either timebase T1 or T2. Users U4 and U5 can access and use only timebase T1 for their analyses. The administrator would set up the Users, A-groups and P-groups as follows:

- Users: U1, U2, U3, U4, U5
- A-groups: A1 = only U1;   A2 = U1, U2, U3;   A3 = U1, U4, U5
- P-groups: P1 = only U1;   P2 = U1, U2, U3;   P3 = U1, U4, U5

Thus, only the members of groups A1 and A2 can access timebases T1 and T2. Users U4 and U5 are restricted to accessing only timebase T1. The P-groups will define those functions that the Users are allowed to perform.

Obviously, the assignments of access and privilege can become much more complex than the above example. Any number of access and privilege groups can be set up with an individual being a member of one or many groups, permitting the individual to have certain privileges when dealing with one data set (for example, as a Study Director) but being restricted when dealing with a different data set (for example, as a reviewer).

The administrator handles the function of granting the accesses and privileges to the users, but the actual decision as to which user may be given which access and privileges must be decided by management, usually through a written Standard Operating Procedure (SOP).

Management determines the system administrator, who may also be a user or may be a member of a separate, independent department. Each institution must decide upon the level of security that is needed to keep data protected from unauthorized access and undocumented changes. In a small research group, most workers might have the privilege of collecting data on all workstations and timebases and to access, view and manipulate all data. In a large organization, only those persons directly involved with a study might have access to a particular workstation, timebase and data. These decisions must lie with the organization and should be described in its SOPs.

An approach that has worked well at the author's organization has been to restrict access to a timebase to the Study Director and his or her supervisor. The Study Director, under GLPS, has ultimate control over the study and then makes the decision regarding which other users (management, technicians, etc.) should be granted access to timebases used in the study. He or she also decides what privileges, if any, each of these persons should be granted. For example, the supervisor might be granted full access and privilege to view all files and data associated with a study, but have no privileges allowing manipulation of the data or preparation of a report. Such a situation would guarantee that the Study Director has complete control of the study, but still allow the data to be readily available to management. A list of allowed users and associated privileges is prepared and given to the Administrator who make the necessary additions or changes to the CMUSER database. The authorized users and their privileges become a part of the raw data associated with the study, and a list is stored electronically in the audit trail as a Chromeleon file. The audit trail becomes a part of the raw data for the study and is archived as such.

The proper use of an automated data system such as Chromeleon, when combined with a well-managed laboratory environment, can be a substantial aid in maintaining a high level of integrity in the collection, documentation and storage of chromatographic data for FIFRA, TSCA and FDA required studies.

References

1. USEPA, *Good Laboratory Practice Standards*, Federal Code of Regulations, 40 CFR Part 160, Federal Register, August 16, 1989.
2. USEPA Directive 2185, Good Automated Laboratory Practices, Office of Information Resource Management, Research Triangle Park, NC, 1995.

Chapter 11

User Testing Strategy for Millennium[32] Chromatography Software Validation

Timothy J. Stachoviak

Information Technology, Covance Laboratories, Inc., Mail Code 31, 3301 Kinsman Boulevard, Madison, WI 53704

> The complexity of chromatography acquisition and processing software makes exhaustive testing impractical. Proving the system meets the user-specified requirements is a manageable alternative to testing every combination of features of the application. Necessary data collection, processing, and presentation features can be extensively tested while instrument control, spectra libraries, and other unused features can be omitted.

User acceptance testing of scientific software is most frequently performed for the purpose of satisfying a regulatory mandate. A more laudable and practical motive is for assurance that the results produced by the system are valid. Moreover, the testing can identify situations where the software does not meet the needs of the user.

The form that the testing takes can be a time-consuming exhaustive trial of each software feature. A viable alternative is to selectively test the features that are needed in the installation. Selective testing will require the users and testers to correlate the software features with laboratory functions and requirements.

Choices must also be made on how to test the selected features. One approach is to set up cases where the systems would be expected to fail because a parameter is specified outside of the application limitations. This approach requires a significant amount of planing. In addition, it is redundant to unit

© 2002 American Chemical Society

testing performed by the software developers. A more reasonable approach is to use data and parameters that match the actual workings of the laboratory. The immediate benefit is that on the completion of testing the system will have been shown to work in the intended environment. The disadvantage that must be kept in mind is that as the environment changes one cannot assume that the system is going to work as expected. Additional testing may be required as new needs are identified.

User acceptance testing is only one of several types of testing in the system validation process. There is testing done by software and hardware developers, testing that software is installed correctly, testing that hardware is functioning, and testing that other computer components (e.g. Printers, databases, networks, workstations, laboratory equipment) are performing as expected. User acceptance testing should attempt not to duplicate these efforts. Especially in the case of unit testing by the developers, incorporating the results of other testing will save time and effort.

Testing must be completed before the system can be used in a regulated environment, but additional benefits are available if the testing is started before a commitment is made to a particular system. The system requirements will necessarily be defined early in the process. The requirements will be refined further as systems are less formally tested. The result should be more objective information to justify the final system selection.

Documents

A chromatography data acquisition system should provide the basic functions of system security, acquisition configuration, chromatography acquisition, chromatography processing, and result reporting. The system requirement specification (SRS) is a listing of the details of these basic functions. The users of the software provide the requirements to the administrator of the system who incorporates them into the SRS. The SRS is the key document in successful User Acceptance Testing. The SRS provides a framework on which to build the test plan. If developed early in the process the SRS can be used as a system "shopping list." The SRS should be independent of the system being tested.

The user acceptance test plan is a specification of what will be done. It will specify necessary setup conditions and any required input data. The plan contains the test log, which gives instructions for the tester to follow, defines expected outcomes, and provides for the tester to record actual outcomes and observations. The plan must be rewritten to address the specific system being tested.

The user acceptance test report addresses test failures and deviations from the plan. The administrator assesses the impact of each item, makes recommendations for use, and determines what future action is necessary. It is acceptable and not uncommon to have test failures. The system can still be used. The other key component of the report is a table listing every item in the SRS with a reference to where it was tested.

Actual Implementation

Environment: Four types of laboratories in a Contract Research Organization (CRO) regulated by FDA 21CFR58 (GLP) and 21CFR11 (Electronic Records and Signatures) and by ISO9000.

Existing Chromatography Systems: Microsoft DOS and 16-bit MS-Windows applications, including Millennium 2020.

Laboratory, quality assurance, and information technology users provided requirements to the Chromatography Data Acquisition System (DAS) team. The team generated a System Requirement Specification (SRS) and solicited proposals from vendors. Millennium[32] was chosen before User Acceptance Testing was begun.

The Application Administrator (AA) performed a vendor audit and obtained unit test results.

The AA prepared a test plan based on the requirements of the SRS, but omitting the features tested by the vendor. The test plan comprised modules that tested the basic functions of security, configuration, acquisition, and processing and reporting by simulation of laboratory operations. Additional modules were needed to specifically test security and privilege assignment because they are not encountered in day-to-day laboratory operation. A final module acquired data in parallel with the existing data acquisition systems to prove the results were the same. The traceability matrix was created to document that each item of the SRS was verified in a validation test.

Information technology personnel executed the tests in a validation environment designed to simulate the production environment as accurately as possible. Execution of the testing was recorded by Lotus ScreenCam and saved as executable files that can be played back to review the workstation display of the actual test.

Evaluation

The first success of the plan was to identify a memory leak in Millennium^{32} version 3.0. The decision was made to delay implementation until the leak was fixed.

Testing on Millennium^{32}, version 3.05, detected additional program errors. ASCII formatted result table exports contained unexpected line breaks. Some user privileges that appeared to be independently settable depended on the setting of other privileges. Peak-fitting algorithms projected a higher than actual peak height for square waves. The Copy-to-Project dialog allowed users to see project names to which they did not have privilege.

Some deficiencies were discovered after release to production. There was no way to assign a read-only privilege to view sample histories for auditors. It was possible to contrive to save a quantitated peak without saving the calibration that was used for the quantitation. When assigning users to groups the system would incorrectly match to a longer name that had the same initial substring. In some cases, curve statistics were not replaced when curves were recalculated. In some cases, updating custom calculation values required reprocessing of the chromatograms. Manually identifying an internal standard peak could lock-up the workstation. The converter for restoration of version 2 projects failed on a few projects.

The evaluation of the success of the validation approach must answer the question of whether taking another approach would have caught the bugs and that the other approach would have not missed other bugs. The answer is most likely that no one would have insight to test for these cases. Most of them require a precise sequence of events to be followed to create the error. Others rely on the chance selection of the input data, so nothing short of infinite testing on all data would guarantee success.

A more pragmatic measure of success is that several client audits of the Millennium^{32} validation have found it sound.

Chapter 12

Auditing Electronically Captured Analytical Chemistry Data

Mary E. Lynn

The Windward Company, 4821 105th Avenue, N. W., Gig Harbor, WA 98335

> Auditing data captured electronically has produced new challenges for Quality Assurance (QA) personnel. In general, auditing electronically captured data during studies conducted in compliance with Good Laboratory Practices (GLP) should be approached no differently than performing any other data audit. However, there are additional considerations when auditing data generated electronically. This chapter discusses these additional considerations for QA personnel. Additionally, suggestions are provided for the QA auditor when conducting in-process inspections, data/report audits, and facility inspections and monitoring records retention.

Auditing data captured electronically has produced new challenges for Quality Assurance (QA) personnel. In general, auditing electronically captured data during studies conducted in compliance with Good Laboratory Practices (GLP) should be approached no differently than performing any other data audit. However, there are additional considerations when auditing data generated electronically.

The current Environmental Protection Agency (EPA) FIFRA GLPs (1) discuss data captured electronically in the context of data recording. Additionally, computers and other electronic equipment that is used for the "generation, measurement or assessment of data" should be considered "equipment". The proposed EPA GLPs (2) also provide additional language for assuring the integrity of data from computers and other automated laboratory systems. The proposed regulation indicates that "the integrity of data from computers, data processors, and automated laboratory procedures involved in the collection, generation or measurement of data shall be ensured through appropriate validation processes, maintenance procedures, disaster recovery, and security measures."

There are a number of areas to consider when auditing data generated electronically including system validation, security, data audit trail, and retention of records. Documents are available that provide suggestions and guidance in this area and with these concepts. The following documents can be useful for QA and technical personnel working in this area.

- EPA's Good Automated Laboratory Practices (GALP) with implementation guidance (3)
- Organisation for Economic Cooperation and Development (OECD) Consensus Document on the Application of the GLP's to Computerised Systems (4)
- Computerized Data Systems for Nonclinical Safety Assessment: Current Concepts and Quality Assurance (5)
- Food and Drug Administration (FDA) 21 CFR, Part 11; Electronic Records/Electronic Signature Rule with preamble (6)

The documents listed above are just a few of those available and come from several different sources, not all EPA related. It is interesting to note that many of the available guidance documents and resources for validation and use of computerized systems in a GLP environment are comparable in the information they provide. Therefore, this chapter does not limit its use of references to EPA related materials.

Quality Assurance Considerations

QA personnel required to monitor the GLP compliance of computerized systems should be familiar with and/or receive training on each system that is utilized in electronic data capture for the studies they are to audit. This electronic data capture training could include hands-on training, attending an outside training course and reading available documents about the system (e.g., the validation report, applicable Standard Operating Procedures (SOPs), system user manuals). Additionally, it is desirable for the auditor to have training in the regulatory areas related to computer systems. In some companies, there are specific QA personnel with specialized training within the Quality Assurance Unit (QAU) assigned to audit computerized systems. In smaller companies with limited resources, it may not be practical for this specialization.

QA personnel are required to inspect/audit each study conducted in compliance with GLPs, but the extent to which QA personnel are involved in software development and the validation/verification process varies from company to company. In some companies, there is little or no QA involvement in these processes, while in others QA personnel are involved by performing inspections and audits just as they would during a GLP study. It is generally a more successful process if QA personnel are involved. QA personnel can provide assistance in the area of vendor audits for purchased software or can conduct inspections of in-house software development to assure internal procedures are being followed. QA personnel can also provide inspection support during the validation and verification process by conducting in-process inspections and reviewing the resulting data and validation report for accuracy. During system development and validation, QA personnel may also be utilized for regulatory advice to assure the system will meet government standards. Being involved early in the validation process QA personnel become more familiar with the system(s) that will be used.

The QAU should have written procedures (i.e., SOPs) for the conduct of inspections and audits. These procedures should incorporate any considerations for the QA review of electronic data systems. For example, the QAU SOPs should address the role and responsibilities of the QAU in software development, purchase, and validation activities; QAU in-process audit procedures for data collected on-line; the procedure for on-line review of data (e.g., what will be verified, how much data will be reviewed); and the procedure for report audits using on-line data.

In-process Inspections

During the conduct of chemistry analysis in-process inspections, the data collection practices, including the data capture system, should be inspected in addition to observing the procedure being performed. The QA auditor should review the protocol and applicable SOPs for data collection practices as well as the procedure prior to conducting the in-process inspection.

The auditor can observe several items related to data collection practices during in-process inspections. The auditor should observe if the protocol and applicable SOPs are being followed with respect to the procedure and data collection practices; if the appropriate security procedures are being utilized; and if changes are noted appropriately (e.g., the original entry is available, the reason for the change is documented, etc.). Additional items to look for include computers left unattended without the user logging out, user name and password posted in the laboratory, and group use of user name and password.

Additional items that can be reviewed during the in-process inspection include personnel training records and equipment records. The personnel training records should be checked to verify not only that the individual(s) observed during the inspection has the appropriate training in the technique, but also to verify that training on the computer system has been documented.

Equipment records should be reviewed for the equipment used during the in-process inspection. These records generally include documentation of maintenance and, if applicable, calibration. Each computerized data acquisition system should have records documenting maintenance. Maintenance for a computer system should include hardware as well as software. If a controlled change was made (e.g., software upgrade), the records should indicate what was changed, why it was changed, who made the change, when it was made, and was the system revalidated after the change. An authorization procedure should be in place for approving controlled changes and this authorization should be documented.

Computer system maintenance records should also record unanticipated events, what corrective action was taken, who performed the corrective action and when. These events might include system crashes, date/time changes after a power failure, etc.

Data/Report Audits

QA's responsibility for the report is to "review the final study report to assure that such report accurately describes the methods and standard operating procedures, and that the reported results accurately reflect the raw data of the study."(1) To adequately perform a data/report audit, QA personnel need direct access to the on-line data. Access for QA personnel should be in the form of read-only access.

The amount of data audited and how the data points are chosen for audit should be specified in the QAU SOPs. An auditor may choose to perform more thorough and more frequent audits on a recently validated system. The validation report can be used to assist in determining what and how much to audit. For example, if data summary printouts from the chromatographic computer system are used in the report, it is important to review the validation report to verify that this function was tested during validation. If this portion of the computer software was successfully validated, it may be sufficient to verify a few values on each table in the report.

QA personnel should also review data on-line. Data changes should be reviewed to ensure the audit trail was appropriately maintained (e.g., the original and changed data are both available, the date of the change was recorded, the reason for the change was recorded, and the person responsible for the change was identified). No data should be overwritten. For example, reintegrated or recalculated data should not overwrite the original data.

The on-line review could also include items such as tracking several samples through the system and a check of the calibration and integration parameters (meta data). If data were reintegrated, all integration parameters should be saved.

If spreadsheets or statistical packages are used, these should be included in the audit. Items such as input values and equations or routines used should be verified during the data audit. This type of software should also be included in the validation program to assure it is providing the correct output values.

Records Retention

QA's responsibility during GLP studies should not be limited to in-process inspections and data and report audits. The QAU should also review the procedures for storing and archiving electronic data. This review should include assuring back-up and archiving procedures were performed as specified in the SOPs; reviewing the records for any instances of equipment failure and assuring contingency plans were followed; and reviewing to assure long-term storage procedures are followed.

Archived electronic data should be treated no differently than archived paper data. Data should be archived at the completion of the study, an archivist should be assigned, access should be limited to authorized personnel, and the material should be indexed to permit expedient retrieval. Depending on the media used for storage, it may be necessary to provide an area within the facility with specific environmental controls needed to maintain the integrity of electronic data. This should be specified in the data storage SOPs.

Facility Inspections

Computerized systems should be included during facility inspections of analytical laboratories. Items discussed previously, such as, computer maintenance records and personnel training records, can be reviewed more thoroughly during the facility inspection.

In addition to the personnel training records for laboratory technical staff, training records should be reviewed for staff involved in the software development, validation and computer maintenance. These personnel may reside in a separate computer functional group or may be part of the analytical laboratory group. In either case, their training records should contain documentation of education, experience and training to support the duties they perform. Depending on the function performed, this training should include training in GLPs and other regulations or guidance as applicable.

The facility inspection should include a review to assure applicable computer systems and software have been validated or tested as needed. There may be differences between how network systems are validated in contrast to stand-alone systems. The records and procedures for the different systems should be reviewed as part of the facility inspection. If validation reports have not previously been reviewed by the QAU, the facility inspection may provide an opportunity to review these reports and data.

As discussed earlier, each computer system used to capture data during GLP studies should have records of maintenance. The records should be reviewed to assure they are compliant with the applicable SOPs and GLPs. Consistency of record keeping between systems should also be reviewed. Items such as system to system variation in time/date settings, passwords being changed as required, completeness of documentation of software and/or hardware upgrades for each system, and availability of maintenance records, user manuals and other system documentation in the laboratory can be evaluated during the facility inspection.

References

(1) Environmental Protection Agency. FIFRA Good Laboratory Practice Standards; Final Rule. 40 CFR Part 160. 1989; Federal Register Vol. 54, No. 158.
(2) Environment Protection Agency. Consolidation of Good Laboratory Practice Standards; Proposed Rule. 40 CFR Parts 160, 792, 806. 1999; Federal Register Vol. 64, No. 249.
(3) Environmental Protection Agency. Good Automated Laboratory Practices: Principles and Guidance to Regulations for Ensuring Data Integrity in Automated Laboratory Operations with Implementation Guidance. Office of Information Resources Management. 1995.
(4) Organisation for Economic Cooperation and Development (OECD). The Application of the Principles of GLP to Computerised Systems. 1995;

Environment Monograph No. 116. GLP Consensus Document, Number 10.
(5) Drug Information Association. Computerized Data Systems for Nonclinical Safety Assessment: Current Concepts and Quality Assurance. 1998. Drug Information Association.
(6) Food and Drug Administration. Electronic Records; Electronic Signatures; Final Rule. 21 CFR Part 11. 1997; Federal Register Vol. 62, No. 54.

Chapter 13

Good Laboratory Practice Considerations for Electronic Records

Kendy L. Keatley

Gilead Sciences, Inc., 2860 Wilderness Place, Boulder, CO 80301

GLPs are in place to ensure both data quality and integrity. The use of computerized systems and data exchange between systems external to one another dictates the creation of electronic records. As such, these records are subject to the same GLP parameters as paper records. The data must be accurate, authentic, attributable, current and legible. The regulatory guidelines below, although not all inclusive, address implementation and consistency of electronic data interchange, and promote the integrity of any data created, modified, maintained, archived, retrieved or transmitted via computerized systems.

Both the United States Environmental Protection Agency (EPA) and Food and Drug Administration (FDA) have issued regulatory documents to address electronic reporting to the Agencies. Two comprehensive and useful Electronic Data Interchange (EDI) guidelines in place by the EPA are the *EDI Implementation Guideline* (*1*) and Federal Register Notice Interim Final Notice, *Filing of Electronic Reports via Electronic Data Interchange* (*2*). EPA has also issued the *Cross-Media Electronic Reporting and Record-keeping Rule,* otherwise known as CROMERRR *(3).* Although the proposed rule was signed, it is pending publication in the Federal Register while under review by the Bush Administration. The FDA's *Guidance for Industry, Computerized Systems Used in Clinical Trials (4)* addresses a number of GLP aspects of electronic records applicable to all areas of the GLP arena. The FDA's Electronic Standards for the Transmission of Regulatory Information (ESTRI) Gateway to define strategic plans for electronic submissions to the Agency is currently underway. The Agency already has in place *Guidance for Industry, Providing Regulatory Submissions in Electronic Format – General Considerations (5).* In addition, applicable to the scope of all electronic records and signatures is 21 *CFR* Part 11, FDA's *Final Rule, Electronic Records; Electronic Signatures,* effective August 20, 1997 *(6).*

21 CFR, Part 11; Electronic Records; Electronic Signatures

Before the above cited guidelines can be applied, knowledge of the Final Rule for electronic records and signatures is necessary. Following are the provisions of Subpart B – *Electronic Records* and Subpart C – *Electronic Signatures.*

Subpart B – Electronic Records

(§11.10) Controls for closed systems (environment in which access is controlled by persons responsible for the electronic records). Persons using closed systems to "create, modify, maintain, or transmit electronic records" need to employ the following procedures and controls, and ensure the signer cannot repudiate the records as not genuine.

- Validate the systems.
- Generate accurate and complete copies of records in readable and electronic form subject for inspection, review, and copying.
- Protect records for retrieval during records retention period.
- Ensure limited access to authorized individuals.
- Use secure, computer generated, time stamped audit trails for date and time of operator entry/action to create, modify, or delete records; changes should not obscure the previous record; the audit trail must be retained and available for agency review and copying.
- Use operational system checks to permit sequencing of steps and events.

- Use authority checks to ensure only authorized individuals use/access a system, alter records or electronically sign a record.
- Use device checks to determine the validity of data input source or operational instruction.
- Determine that those who develop, maintain or use the system have the education, training and experience for their assigned tasks.
- Have written procedures in place to hold individuals accountable and responsible for actions under the use of their electronic signature.
- Have controls over the distribution, access and use of documentation used for system operation and maintenance.
- Have time-sequenced development and modification of the system's documentation for revisions and change control procedures for audit trails.

(§11.30) Controls for open systems (environment in which system access is not controlled by those responsible for the electronic records). Persons using open systems need to employ the same procedures and controls outlined in *§11.10*. Additional measures for document encryption and digital signatures (signatures based on cryptographic methods) need to be in place to ensure record "authenticity, integrity, and confidentiality".

(§11.50) Signature manifestations. Signed electronic records need to clearly indicate the printed name of the signer; the date and time when the signature was executed; and, the purpose (review, approval, etc.) with the signature. These items are under the same controls as electronic records.

(§11.70) Signature/record linking. Electronic signatures and handwritten signatures executed to electronic records need to be linked to the respective records to ensure signatures cannot be "excised, copied or otherwise transferred".

Subpart C – Electronic Signatures

(§11.100) General requirements.

- The electronic signature is to be unique and not reused or assigned again.
- The organization must identify the individual before using the signature.
- The person is to certify to the agency that the signature used on or after August 20, 1997, is the legally binding equivalent of a traditional handwritten signature prior to, or at the time of, such use.

(§11.200) Electronic signature components and controls. Electronic signatures not based on biometrics (method of identity based on measurement of physical feature or repeatable action that is measurable) require at least two

distinct "identification components", such as an ID code and password. When a series of signatures is used during a continuous period, the first executed signature must contain all of the components, while subsequent signings need only include one of the components. Signings that are not performed during a continuous period must include all of the components with each signature. Biometric signatures should be designed to exclude use by another.

(§11.300) Controls for identification codes/passwords. Controls need to be implemented to ensure the security and integrity of identification codes and passwords that include the following.

- Ensure that the combination used is unique.
- Ensure ID codes/passwords are periodically checked, recalled or revised.
- Have rigorous "loss management" procedures in place to de-authorize lost, stolen, missing or otherwise compromised signatures and re-issue the signature.
- Use safeguards to prevent unauthorized use, and detect and immediately report these attempts to the system security unit and organizational management.
- Test devices that bear or generate ID codes/passwords, both initially and periodically, to ensure that they still function and have not been altered.

Review of FDA's Guidance for Industry, Computerized Systems Used in Clinical Trials

Computerized Systems Used in Clinical Trials addresses various aspects of electronic records and the requirements of the Electronic Records/Electronic Signatures Rule. The Guidance outlines measures to ensure that the fundamental elements of data quality, that is, that the data are "attributable, original, accurate, contemporaneous, and legible", are met where computerized systems are being used. Although the Guidance was written for clinical trials, the principles outlined for electronic records and electronic signatures are applicable to any data where computerized systems are being used. An *electronic record* is defined as "any combination of text, graphics, data, audio, pictorial, or any other information representation in digital form that is created, modified, maintained, archived, retrieved, or distributed by a computer system". An *electronic signature* is defined as "a computer data compilation of any symbol or series of symbols, executed, adopted, or authorized by an individual to be the legally binding equivalent of the individual's handwritten signature". Following are those aspects of the Guidance specific to electronic records and electronic signatures.

General Principles

- The study protocol should define at which steps the computerized system will be used to create, modify, archive, retrieve, or transmit data.
- There should be documentation that identifies both the software and hardware used; the documentation should be retained as part of the study records.
- The original documents or records, also known as source documents, should be retained for reconstruction and evaluation.
- Original observations entered directly into a computer create an electronic record which is the source document.
- The design of the system should ensure that regulatory requirements for record keeping and record retention are met.
- Changes to records should not obscure the original and should indicate that a change was made; there should be a means to locate and read the prior information.
- Changes to data should delineate whom, when, and why the changes were made.
- Computer systems should be designed to meet the specified protocol requirements and preclude errors in data creation, modification, maintenance, archival, retrieval, or transmission.
- Security measures are needed to prevent unauthorized access to the data and the system.

Standard Operating Procedures

SOPs for the use of computerized systems should include (but are not limited to) the following:

- System setup and installation.
- Data collection and handling.
- System maintenance.
- Data backup, recovery and contingency plans.
- Security.
- Change control.

Data Entry

Electronic Signatures

Individuals with authority for data entry need to have established electronic signatures in the form of ID codes/passwords or biometric signatures. The data entry is to be attributable to the individual making the entry including changes made to any entry. The printed name of that person should be displayed on the screen at all times to preclude data entry by someone else and ensure the authority of the individual making the entries. At no time should a system be logged on to providing access by another individual. In leaving a workstation, the person should log off; the system may be designed to do an automatic log off after a designated time period. For absences during short periods, the system design should provide for unauthorized access. At established intervals, passwords and other access protections should be changed.

Audit Trails

The Electronics Records/Electronic Signatures Rule (*21 CFR 11.10(e)*) dictates that electronic record systems maintain an audit trail to ensure the authenticity, integrity, and as appropriate, the confidentiality of those records. As such, individuals must use secure, computer generated, time stamped audit trails to "independently" record the date and time of operator entries that create, change or delete records. The audit trail should be designed to preclude modification of the audit trail by that individual. The audit trail must be incremental and chronological, and is subject to the required record retention period of the study records.

Date/Time Stamps

Only authorized personnel should be able to change a date/time stamp, and that action should be well documented. Measures also need to be taken to ensure the correctness of the stamp. Date/time stamps should include the year, month, day, hour, and minute.

System Features

Systems used for data (direct) entry should ensure the quality of data collection. Measures (i.e., flags, prompts or other "help features") for consistency of use, alerts for unacceptable ranges, and annotations are essential.

Retrieval of Data

Systems for data entry should include features for inspection and review of the data. "Data tags" should be used to distinguish changes or deletions indicated in the audit trail. The ability to retrieve the data is a requirement even under circumstances where the system has been updated. This may have to be accomplished by maintaining support of older systems or by transcribing data to newer systems. If transcribed, the transcription process needs to be validated and complete copies of the study data and any collateral information should be generated.

Reconstruction of Study

Not only the data, but also how the data were obtained or managed needs to be indicated to reconstruct a study. As such, all versions of the software applications, software development tools for processing data and the operating systems must be retained, along with the ability to run the software.

Security

Physical Security

External safeguards need to be in place to ensure limited physical access to computerized systems, and ensure that only authorized personnel have access. Personnel should be fully aware of the security system in place. These security measures should be outlined in SOPs.

Logical Security

The logical security of a computerized system should address the internal safeguards of the system and how access to the data is limited. These safeguards should outline how data access is restricted through use of the software, log on, security procedure and audit trail. A cumulative record accessible with the system should be kept indicating the names of authorized personnel, their titles and their access privileges. If a computerized system is used for other purposes, efforts need to be made to preclude compromising the data through interaction with other software. The system should be re-evaluated if "any" software changes are made to determine changes on the logical security. In addition, controls for computer viruses should be used.

System Dependability

The dependability of a computerized system relies upon documentation that fully describes the hardware, software and physical environment (systems documentation). Requirements for established "completeness, accuracy, reliability and consistent intended performance" should also be met.

Software Validation

"Off the shelf" software may or may not have design level validation (software validation that takes place in parts in the software life cycle before delivered to the end user) from the vendor. Under these circumstances, functional testing of the software must be performed to determine the limitations, problems and defect corrections of the software.

Software validation must be performed and have documentation in place for the following.

- Design specification describing both what the software is intended to do and how.
- Written test plan for structural and functional analysis.
- Test results and evaluation that demonstrates the design specification has been met.
- Written procedures to evaluate, test and re-validate changes in the software, equipment or component replacement or new instrumentation.
- Documentation of all changes.

The FDA has in place *Draft Guidance, General Principles of Software Validation, Version 1.1* (7). Although the Guidance is meant to address medical device software, it is a useful guide for validation of any software either developed in house or purchased off the shelf.

System Controls

These controls include software version control, contingency plans, and backup and recovery. Software version control should ensure that the software used as stated in the systems documentation is the version used for the data collection. Written procedures should be in place to have a contingency plan in the event of a system failure. SOPs need to be in place to fully outline procedures to prevent the loss of data and to address backups of the data. The backups need to be stored in a secure location. Backup and recovery logs should be maintained to assess, as appropriate, any loss of data.

Training of Personnel

Any individual entering or processing data should have the "education, training and experience", or a combination thereof, to do so. Training should be provided not only for specific operations, but also on a continuing basis, as needed, for familiarity with any changes in operation. These requirements are to be documented.

Records Inspection

The computerized system must be able at all times to generate accurate and complete copies of records in both human readable form and electronic form subject to inspection, review or copying.

Certification of Electronic Signatures

The Electronic Records/Electronic Signatures Rule requires certification that the electronic signature of an individual is the legally binding equivalent of their handwritten signature (*21 CFR 11.100 (c)*). The certification is to be submitted in paper form with a handwritten signature.

Overview of Electronic Submissions

EPA Submissions

The EPA is currently in the implementation phase of the legal framework for a rule to address electronic submissions to the Agency. The rule titled the *Cross-Media Electronic Reporting and Record-keeping Rule* (CROMERRR), when codified, will be used for electronic reporting to the Agency.

Current guidance for electronic submissions to EPA is in place under the *Filing of Electronic Reports via Electronic Data Interchange (EDI)*. The policy includes general information for reporting of regulatory, compliance or informational purposes via EDI. The policy is meant to streamline and simplify legally admissible regulatory reporting to the Agency, and is meant to promote consistency in implementing EDI. It is important to note that electronic submissions must meet the same legal signature/certification requirements and any other regulatory requirements of paper submissions. Following are highlights of the policy.

EDI is defined as "the transmission, in a standard syntax, of unambiguous information between computers that may belong to organizations completely external to each other". Currently EDI is based on standard formats and protocols under the American National Standards Institute (ANSI) Accredited Standards Committee (ASC) x 12. EDI reporting is subject to the Agency's Terms and Conditions Agreement (TCA) which must be signed by the submitter before electronic reporting will be accepted. The TCA is used to certify and/or authenticate the submitter of reports to the Agency. Although this policy only provides a generic TCA model, there are program specific TCAs that should be obtained from the Agency.

Once a TCA has been submitted, EPA issues a Personal Identification Number (PIN). The assignment of a PIN is meant to ensure the integrity and authenticity of electronically submitted reports. The submitter's PIN must be included in all reports and is deemed to indicate authenticity. Also, responsibility and accountability for the PIN is directly linked to the individual assigned that PIN. That individual is responsible for both the accuracy and authenticity of the information submitted. It is the corporate officer of the submitter that must identify authorized personnel who may use PINs. It is also the submitter's responsibility to immediately notify EPA of personnel changes or if a PIN has been comprised in any way. The submitter must institute and maintain security procedures to ensure the security of assigned PINs. Records must be retained for the assignment and revocation of PINs.

All record keeping requirements in existing regulations are applicable to electronic reporting. Submitters must retain records to ensure the authenticity, completeness, accuracy and integrity of electronic transmissions. Those records should create an audit trail for both the creation and submission of electronic transmissions. A Transmission Log is required to be kept for all parties using EDI. The Transmission Log should include the date, time, destination address and telephone number, and copy of the transmitted file. The documentation should also include who had access to the system during the creation and transmission of the files. The Log is to be retained without modification. A qualified individual with appropriate authority should be designated as responsible for the Log.

The *EPA EDI Implementation Guideline* is a detailed reference for EDI implementation. For updates, EPA has available *Electronic Reporting at EPA: Electronic Commerce/Electronic Data Interchange (EC/EDI)* at URL http://www.epa.gov/oppeedi1.

FDA Submissions

The FDA has a number of aggressive initiatives underway for implementing electronic submission and review systems. The Electronic Standards (for the) Transmission (of) Regulatory Information, also known as ESTRI is FDA's "Gateway" that allows electronic filing of regulatory information. The Gateway applies a core set of open standards incorporating the International Committee for Hamonisation (ICH) M2 standardization efforts. More detailed information on the Gateway can be found at FDA's Frequently Asked Questions site at URL http://www.fda.gov/oc/electronicsubmissions/interfaq.htm.

FDA's *Guidance for Industry, Providing Regulatory Submissions in Electronic Format – General Considerations*, outlines general issues common to all types of electronic submissions. It is one of a series of guidance documents and addresses submissions to the Center for Drug Evaluation and Research (CDER) and the Center for Biologics Evaluation and Research (CBER).

Currently FDA recommends submission of Portable Document Format (PDF) files, which is a published file format, created by Adobe Systems Incorporated (www.adobe.com). The file format is intended to adhere to the requirements of the Electronic Records/Electronic Signatures Rule that includes the following.

- Enable the user to easily view a legible copy of the information.
- Enable the user to print the document, page by page, maintaining fonts, special orientations, table formats and page numbers.
- Include a well structured table of contents.
- Allow the user to copy text and images electronically into common word processing applications.

Some general aspects include the use of Times New Roman 12 point font, 8 1/2 by 11 inch page size with a 1 inch margin on all sides, landscape orientation, and avoiding the use of scanned documents. If scanned documents are used, the submitter needs to assure the resolution is such to allow the documents to be readable both on a screen and on paper. It is also suggested to avoid any applications that result in increased file size. The files should not include any security settings or passwords. The Agency should be able to read the file with Adobe Acrobat version 3.0 without the use of plug-ins. Although procedures for archiving documents with electronic signatures are being developed, currently, any documents requiring original signatures must also be submitted.

For datasets provided in electronic format, the FDA is currently able to accept datasets in SAS System XPORT transport format (Version 5 SAS transport file). SAS XPORT is an open format published by the SAS Institute (www.sas.com/fda-esub). SAS transport files should not be compressed and should be organized so that their size is no more 25 MB per file. A single transport file should be used for each dataset.

Recommended electronic media for submission include 3.5 floppy disks, CD-ROMs or digital tape and should be sent directly to the appropriate FDA Center. All electronic media should be adequately secured and designated as "Electronic Regulatory Submission for Archive". The media should be labeled with the following.

- Submission identifier.
- Proprietary and generic name.

- Company name.
- Submission serial number.
- Submission date (DD-MM-YYYY).
- Disk/CD-ROM/tape number identifying the total number.

The FDA has established a public docket number 92S-0251 (8) that lists the FDA Centers that are prepared to receive regulatory submissions and specific records that can be accepted. The FDA intends to establish a series of guidance documents on electronic regulatory submissions for the following: New Drug Application (NDAs) to CDER; Marketing Applications to CBER; Abbreviated New Drug Applications (ANDAs); Postmarketing Safety Reports; Investigational New Drug Applications (INDs); Annual reports; Drug Master File (DMFs); Launch Material; and Advertising. Updated information on electronic submissions can be found at URL http://www.fda.gov/ora/compliance_ref/part11 or by e-mailing to edigateway@oc.fda.gov.

References

1. EPA, *EDI Implementation Guideline*, Draft of September 23, 1994 and October 18, 1994, URL http://www.epa.gov/oppeedi1/guidelines/general.pdf.
2. EPA, *Notice of Agency's General Policy for Accepting Filing of Environmental Reports via Electronic Data Interchange (EDI)*, Interim final notice. Federal Register Vol. 61, No. 172, 46684, September 4, 1996, URL http://www.epa.gov/oppeedi1/edipolic.htm.
3. EPA, CROMERRR, *Establishment of Electronic Reporting; Electronic Records*, January 19, 2001, URL http://www.epa.gov/cdx/cromerr_rule.pdf.
4. FDA, *Guidance for Industry, Computerized Systems Used in Clinical Trials*, April 1999, URL http://www.fda.gov/ora/compliance_ref/bimo/ffinalcct.htm.
5. FDA, Guidance for Industry, Providing Regulatory Submissions in Electronic Format – General Considerations, January 1999, URL http://available at www.fda.gov/cder/guidance/index.htm.
6. FDA, 21 CFR Part 11, *Electronic Records; Electronic Signatures*; Final rule. Federal Register Vol. 62, No. 54, 13429, March 20, 1997, URL http://www.fda.gov/ora/compliance_ref/part11/default.htm.
7. FDA, *Draft Guidance, General Principles of Software Validation, Version 1.1*, 1997, URL http://www.fda.gov/cdrh/comps/swareval.html.
8. Public Docket Number 92S-0251, URL http://www.fda.gov/ohrms/dockets/dockets/92s0251/92s0251.htm.

Chapter 14

Metrology: A Tool and Approach to Ensure Data Quality

Gail E. Schneiders[1], John C. Brown[2], and Joseph Manalo[2]

[1]DuPont Crop Protection, Newark, DE 19714
[2]DuPont Pharmaceutical Company, Route 141 and Henry Clay Road, Wilmington, DE 19880–0353

The computer has become a major tool in the scientific laboratory for the capture, manipulation, transfer and storage of data. As a result, the main focus on data quality has shifted from the instruments that generate the data to these electronic systems, often neglecting the fact that the data are only as accurate as the instrument measurements. Metrology, the science of measurement and calibration, can provide a disciplined approach for ensuring the accuracy of scientific measurement, and a metrology program should be included as part of an organization's scientific best practices. To achieve this goal a metrology program needs to consist of four main functions. These functions include: 1) an inventory tracking system; 2) a process for qualifying new instruments or re-qualifying them periodically or when there are changes; 3) a calibration and maintenance program to assure proper function and performance; and 4) a point of control for all critical documents needed to demonstrate that the equipment or systems have been qualified and calibrated, and perform in a way that ensures the quality of the data. The following outlines a framework and components for a metrology program. This program provides the necessary controls to promote a disciplined approach for ensuring measurement accuracy and precision, from the point of instrument installation, throughout its life-cycle, to meet both scientific and regulatory goals.

Introduction

Metrology is defined as: "The Science of Measurement for the determination of conformance to technical requirements including the development of standards and systems for absolute and relative measurements." *(1)* In other words, it is the science used to demonstrate that an instrument performs at a specific level of accuracy and conforms to known standards. Data generated on this system should be reproducible and consistent. A program based on metrology principles can provide an organization with a measure of assurance that the data generated are true and accurate as measured. This means the instrument meets performance standards and contains proper documentation of equipment qualification, calibration and maintenance. Including computer system validation in the program provides a means to ensure the data integrity throughout its life-cycle. A good metrology program with outstanding documentation practices and controls can meet the compliance needs of current Good Manufacturing Practices (cGMP), Good Laboratory Practices (GLP) or International Organization of Standardization (ISO) Guide 25 standards.

There are many reasons an organization may consider setting up a metrology program to ensure data quality. The first question an organization should ask is who are the stakeholders and how do they use the data provided. For instance, the equipment or system user wants to generate high quality work, with reduced equipment system failures and elimination of the need to conduct repeat analyses. The business organization wants confidence in the data, reduced costs, increased capacity, consistent practices across the organization, reduced product recall, while meeting customer needs. The customer wants the assurance that the product meets their specifications. For many organizations there are regulatory requirements regarding data validity and compliance with documented procedures. Non-compliance with these requirements can result in regulatory actions against an organization.

To satisfy the Environmental Protection Agency (EPA) *(2)* and Food and Drug Agency (FDA) good laboratory and manufacturing practices regulatory requirements, as well as international regulatory requirements, instruments need to be adequately tested, calibrated and/or standardized according to documented procedures. Current EPA FIFRA regulations state: "Equipment used in the generation, measurement, or assessment of data and equipment used for facility environmental control shall be of appropriate design and capacity to function according to the protocol and shall be suitably located for operation, inspection, cleaning, and maintenance." *(3)* The EPA 1999 draft GLP regulations, which consolidate the 1989 FIFRA and TSCA GLPs *(4)*, are more specific on the requirements for ensuring electronic data integrity: "The integrity of data from computers, data processors, and automated laboratory procedures involved in the collection, generation, or measurement of data shall be ensured through appropriate *validation processes, maintenance procedures, disaster recovery and security measures*." *(5)*

Elements of a Metrology Program

A metrology program, capable of satisfying its various stake holders, is composed of multiple elements: an accurate inventory and tracking system (database) containing information on individual components of instruments or systems; a process for qualifying instruments when purchased or when a component is upgraded, a calibration and maintenance program, and an effective record keeping system (Figure 1). The program should be defined in a metrology or qualification program description, containing standard operating procedures (SOPs) for each step of the process, and have personnel with appropriate training for their responsibilities.

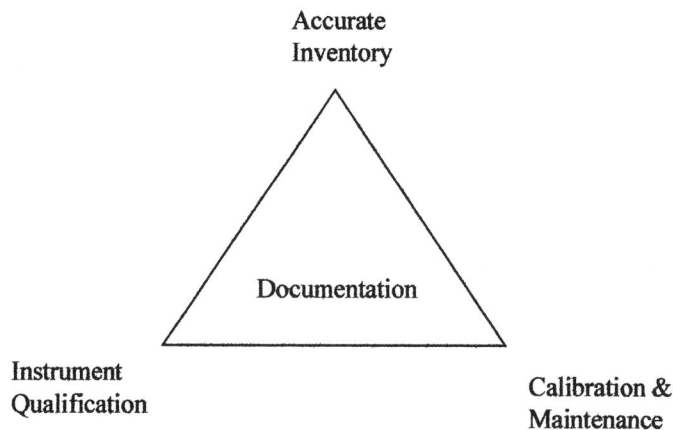

Figure 1 Metrology Program – Ensuring Data Quality

Inventory Management

Key to ensuring data quality is being able to verify that the instruments and systems were performing accurately at the time of data generation. In order to do this, an organization needs to know which instruments and systems they own, as well as their historical and current state of calibration and maintenance. The simplest way to do this is to purchase metrology database software. For organizations with only a small number of instruments and systems, the database could be as simple as an Excel spreadsheet. These metrology or Excel databases should contain appropriate data fields, such as, system and individual component identification, manufacturer, model, serial number, description, location, and custodian. In addition, critical performance parameters, such as, range, resolution and user requirements, can help identify which equipment should be used for specific measurements. The status of the instrument and components (qualified - active, out of tolerance - locked out, retired) and calibration dates/schedule can also be tracked. A sample instrument information form is shown in Figure 2.

A good metrology database will provide automatic reminders when the equipment is ready for re-calibration or maintenance. A person responsible for taking inventories and maintaining the database is necessary for an inventory management system to be effective.

Additional benefits of an instrument inventory include the ability to shift workloads among instruments, facilitate scheduling of calibration and preventive maintenance, and leverage calibration and preventive maintenance contracts. Developing data on the historical performance of an instrument and components facilitates equipment renewal assessments to either purchase additional equipment because of workload demand or replace older equipment that is showing repeat failures or out-of-tolerance responses. Information on the version of software being used by the equipment facilitates routine upgrades and software validation of the equipment operating systems.

Qualifying Instruments

The life cycle of an instrument or system consists of many stages (Figure 3). It begins with the identification of a need to generate specific data with defined performance parameters; then, identification of vendors whose products meet those performance specifications, followed by purchase, installation, qualification, daily use, maintenance, failures and repairs, and eventually retirement and replacement.

The first step in an instrument life-cycle is identifying an instrument need that is currently not being met. A detailed list of specific design parameters

Component ID: _____ System ID: _____
Custodian to fill out:
Description: _____
Manufacturer: _____ Purchase order number: _____
Model: _____ Acquisition date: _____
Serial number: _____ Location: _____
Custodian: _____ Manuals: _____
Metrologist to fill out:
Type: _____ Usage: CRITICAL   REFERENCE   NON-CRITIC
Status: ACTIVE   inactive   retired   Status date: _____
Department: _____ Logbook: _____

CRITICAL INSTRUMENT CALIBRATION INFORMATION:
Range & Resolution: _____
User Requirements: _____
Calibrator: _____ Calibration SOP: _____
Calibration Points & Tolerances: _____
Interval type: MONTH   END OF MONTH   SPECIFIC MONTH   N/A   Number: _____
Basis for Date Calculation: CURRENT DUE DATE   DATE OF EVENT   N/A
Comments: _____
_____

Figure 2. Instrument User Information Form.

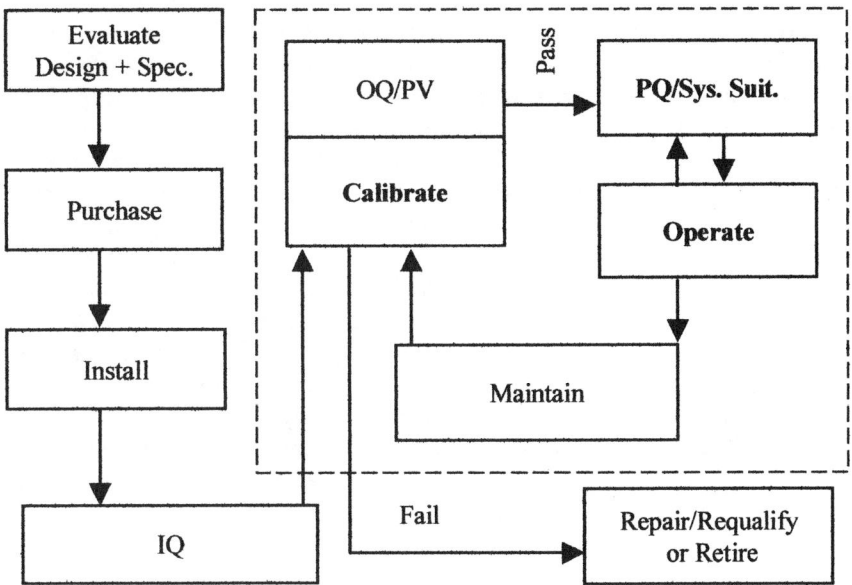

Figure 3. Instrument Qualification.

and specifications needs to be created. The purchase criteria or design specifications for the instrument should include the intended use of the instrument, critical parameters, specific ranges, disposition of raw data, and the user needs. These criteria are then compared to the vendor specifications to ensure that the proposed instrument has the functional range to meet these identified needs.

Once the instrument has arrived, it should be set up and its performance qualified against both the manufacturers's specifications and the purchase criteria. Draft revisions to the ISO 25 guideline state: "Equipment shall be capable of achieving the accuracy required and shall comply with specifications relevant to the tests and calibrations concerned. New equipment shall be checked against the purchase order to establish that it meets the laboratory's specification requirements, complies with the relevant standard specifications, and is calibrated and/or verified before use." (6) The Installation Qualification (IQ) demonstrates the equipment/system has been installed correctly at the user site according to vendor standards. The vendor should install the equipment to demonstrate to the buyer that all the components are operating appropriately. The qualification process includes appropriate documentation of the system components, physical installation and hook-up, and a performance check to verify the individual components operate and can communicate with each other. System component information, serial numbers, type of use and user performance requirements should be captured in the metrology database for easy tracking and scheduling of maintenance and/or calibration.

Operation Qualification (OQ) involves verifying the system operates according to the specifications as agreed on between the vendor and purchaser. This should include a test of each critical component and function according to the vendor specifications and user requirements (if different) using specific standards. The Operation Qualification is usually done by the vendor; however, in-house or a qualified third party contractor may be used. The qualification can be conducted on each component of the system or holistically on the entire system. If each component is qualified, then the system has been operationally qualified and is ready for Performance Qualification (PQ). Conversely, if one component does not qualify, the system cannot be qualified! If the instrument contains computer software, then its ability to accurately capture, store, transfer and manipulate the data should be validated at this time. Detailed test scripts and a complete qualification documentation package needs to be generated. Many vendors now supply OQ documentation packages. However, in-house procedures need to be written to address custom specifications. If a component is replaced or upgraded, a new IQ and OQ needs to be conducted on this component. A complete operation qualification is usually not required for the system.

to be conducted on this component. A complete operation qualification is usually not required for the system.

Performance Qualification and System Suitability (SS) demonstrate that the performance specifications of the system meet the user's expectations and needs for a given use. Performance Qualification is often called method validation. Method validation can be a general validation of commonly used parameters or a specific method validation. At a minimum, it should include expected performance and limit/failure testing. Performance Qualification of a method can include demonstration of precision, resolution, separation, recovery, and signal-to-noise ratios. It should be done before putting the instrument into routine use. Performance Qualification may be repeated many times during the life-cycle of the instrument as new methods with different performance criteria are used. Additionally, it should be conducted after routine calibration and maintenance, relocation, repair and component upgrades. The user or a qualified third party contractor can do this qualification. System suitability is performed daily by the user, usually prior to and throughout sample analysis, using specified standards and performance criteria. Again, the performance of the system is documented so that the accuracy of data generated can be verified. Table 1 summarizes the roles and responsibilities of the vendor and purchaser during instrument qualification.

Table 1. Instrument Qualification Roles and Responsibilities

Activities	*Accountability*	*Options*
Design Specifications	User	
Installation Qualification	Vendor	
Inventory Management	Metrologist, User	
Operation Qualification	In-house Vendor	Third Party contractor
Performance Qualification and System Suitability	User	Third Party contractor
Calibration/Maintenance	In-house User	Third Party contractor

Calibration and Maintenance

To facilitate the maintenance of equipment with different performance criteria, written procedures are needed. These procedures serve as a record of the process used to evaluate the systems' performance. The maintenance technician can sign the procedure record on completion of the task, verifying the equipment meets the performance criteria. Equipment may need calibration as part of the maintenance procedure. Maintenance by appropriately trained personnel should be performed at regular intervals, before equipment parts fail. The technician could be any of a number of involved personnel: the person responsible for the metrology program, laboratory personnel who are familiar with the instrumentation, specifically trained maintenance or calibration personnel, third party contractors, or the original equipment manufacturer. Each choice has its benefits, and different scenarios may be used within the same company. If an organization is relatively large, it may make sense to have in-house personnel whose primary job is to maintain and calibrate the equipment. However, for smaller organizations or for specialized equipment, using outside experts may be more cost effective.

Defined maintenance procedures should include model or manufacturer specifics and a list of parts to be inspected, cleaned, lubricated, replaced and/or calibrated. The replacement part numbers, cleaning solutions and lubricants, and calibration standards, along with the manufacture's maintenance procedures to be followed should be specified. Documentation is easily managed by creating a one-page checklist of instructions or performance parameters that can be checked off by the technician as each task is completed. Any issues or comments can be captured directly on the checklist. Provisions for failure or out-of-tolerance notification need to be clearly defined, as the equipment cannot be put back into service until the performance has been verified by conducting a performance qualification.

As part of maintenance, some equipment may also need to be calibrated. Written calibration methods and a report format need to be available at the time of calibration. A calibration technician is responsible for performing and tracking calibrations, and writing calibration reports. The technician's responsibilities may also include handling and re-certification of standards. For quality calibration standards use NIST (National Institute of Standards and Technology) standards or other intrinsic standards of known purity, quality and stability. These standards should have certificates attesting to their performance properties. When an instrument is calibrated, the accuracy ratio of the instrument to the standard is ideally 10:1; however, a 4:1 ratio is acceptable. For example, if a thermometer is accurate to $\pm 1.0°C$ and the temperature standard used to calibrate it is accurate to $\pm 0.25°C$, then the accuracy ratio is 4:1. If the accuracy ratio drops below 4:1, an uncertainty analysis is required. An uncertainty analysis (7) is the probability of false calibration decisions based on the accuracy of the standard relative to the instrument under test. After calibration, the metrologist or other responsible person should review the calibration reports to identify any issues with the equipment that may need further attention.

Documentation

An organized document filing system must be maintained. This could be a paper file, an electronic document file or a mixture of both. The equipment inventory system already contains key information on the components of each system, their performance criteria and maintenance and calibration status. Additionally the qualification process will generate installation and performance documentation. Other documents necessary to demonstrate the quality of the data include standard operating procedures for the qualification procedures, calibration, maintenance, personnel training, etc. If the manufacturer's operating, service or maintenance manuals are used or cited in the operating procedures, copies of these manuals should also be maintained. To facilitate retrieval, documentation should be stored in a central location and be indexed for easy retrieval.

Regular Process Audits

In addition to incorporating the preceding elements into any metrology program, periodic audits of the instrument qualification, calibration and maintenance practices should also occur. This is particularly important for systems generating data that are subject to review by regulatory agencies or certifying organizations. Audits should check the thoroughness and completeness of documentation and procedures, as well as adherence to the procedures. Documentation auditing will include qualification records, calibration records, maintenance records, and training records of those responsible for the organization's metrology program. Additionally, adherence to written procedures and a check against current regulatory standards applying to the organization should be done. These process audits provide a system of checks and balances to help ensure data quality. Two of the most frequent citations by auditing organizations are failure to have SOPs in place and, if in place, failure to follow them.

Metrology Program – Ensuring Data Quality

An organization interested in ensuring data quality from the time of generation of the data through its life-cycle should seriously consider including a metrology program as part of their scientific best practices. Many elements of such a program may already exist within the organization, such as standard operating procedures, certified reference standards, quality assurance/quality control verification, and calibration and instrument maintenance. A formalized metrology program will provide a point of control and standardization of processes that could significantly reduce the cost of generating true and accurate data and result in more satisfied customers.

References

1. Military-Standard (MIL-STD) -11309C PAR 3.1.369
2. The EPA GLP's currently exist in two separate regulations: at 40 CFR Part 160, applicable to Federal Insecticide, Fungicide, and Rodenticide Act (FIFRA): and at 40 CFR Part 792, applicable to the Toxic Substances Control Act (TSCA).
3. USEPA, *Good Laboratory Practice Standards*, Code of Federal Regulations, 40 CFR Part 160, Section 160.61,54, 158, August 17, 1989, 34052.
4. USEPA, Toxic Substances Control Act: Good Laboratory Practice Standards, 40 CFR Part 792, Federal Register Vol. 54, No. 58, August 17, 1989.
5. USEPA, *Consolidation of Good Laboratory Practice Standards* (40 CFR Parts 160 and 792), Federal Register, 64, 249, December 29, 1999, 72972.
6. ISO/IEC Guide 25, General Requirements for the competence of testing and calibration laboratories, draft 4 section 3.4.2, International Organization of Standardization, 1996.
7. The uncertainty analysis formula is: $[a^2 / (a^2 + b^2)] * 100\%$

Chapter 15

Documentation Requirements for the Design of Good Laboratory Practice Software

Robert D. Walla

Astrix Software Technology, Inc., Suite 302, 175 May Street, Edison, NJ 08837

Validation of computerized systems is required by many regulations, quality programs, and company standards. The process of designing software systems that are subjected to validation must be conducted under strict procedures with adequate documentation. The process of designing the system must be well documented and strict controls should be in place to insure the system will ultimately pass validation. The system developer plays a key role in insuring that the system being designed meets minimum documentation standards necessary to satisfy the needs of the validation team. This paper will describe the "best practices" that a software developer should follow to adequately document the design phase of a system that will ultimately be used as a GLP system.

Overview

The term validation has been defined in the literature by many different authors and sources. The most commonly accepted definition of validation can be found in the guideline *General Principles of Validation.* "Establishing documented evidence which provides a high degree of assurance that a specific process will consistently produce a product meeting its predetermined specifications and quality attributes." The approaches that different companies take to satisfy this requirement are varied. This paper will discuss the steps that a software vendor should follow to adequately document a system that will be subjected to validation testing.

Project Assessment and Development

The Product Development Cycle for software that is subject to validation testing is composed of discrete steps that are characterized by rigorous quality control, continuous user feedback, and thorough documentation. The process begins with an assessment of the business needs (Figure 1). The requirements of the system are identified by meeting with the target users. Once the system requirements are known and use cases are designed a technical architecture is proposed. The architecture must employ technology that has a proven and dependable track record. A Project Assessment document is generated, that details the use, benefit, and preliminary budget for the system.

Once the project team has completed the assessment phase, the planning phase will begin. The project team will assemble the following documents: Project plan, quality assurance plan and system test strategy. The complete design of the system will be detailed in the system specification. Once the design is complete and approved the system will transition into production.

Requirements Definition Phase

The developer must meet with the users and content experts to understand the user's needs. A core team of client content and business and information technology leads must be organized to interact with the business assessment team. Typically, a series of meetings will be scheduled to allow the developer to gather the necessary information to complete the clients' needs and objectives.

The requirements will be separated into logical groupings: Functional, technical, reporting, etc. The requirements will also be ranked according to their priority. The developer should utilize a requirements database to manage the user requirements gathered during this phase. A printout of the requirements should be available upon request.

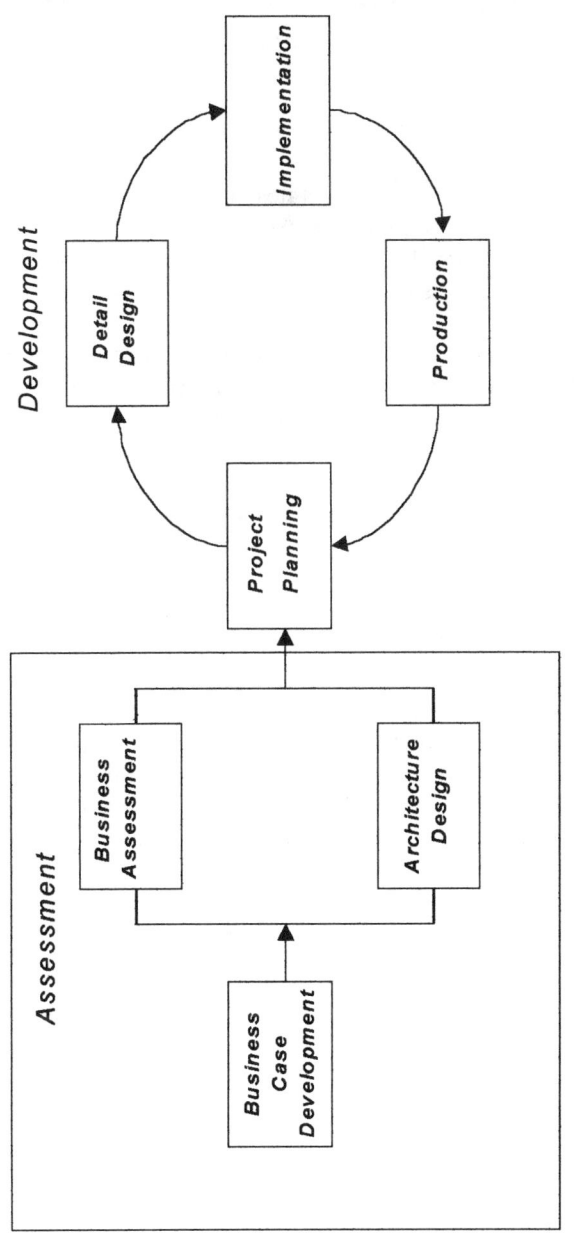

Figure 1. Assessment and Development.

A technical architecture will be proposed after the requirements definition phase is completed.

Technical Architecture Phase

The architecting process incorporates a technical process and an organizational process. The technical process includes steps and heuristics for creating a good architecture. However, a technically good architecture is not sufficient to ensure the successful use of the architecture, and the organizational process is oriented toward ensuring support for, and adoption of, the architecture. The architecture should be documented using adequate tools that show the relationships of the various architectural components and their interaction.

Logical Data Design

After the requirements are gathered and the technical architecture is finalized the database will be designed to accommodate the business needs. A conceptual entity-relationship model shows how the business world views information. It suppresses non-critical details in order to emphasize business rules and user objects. It typically includes only significant entities that have business meaning, along with their relationships. A conceptual model may include a few significant attributes to augment the definition and visualization of entities. A logical entity-relationship model, the blueprint for building an application, is a normalized diagram of data that is used by the business. A logical data model requires a complete scheme of identifiers or candidate keys for unique identification of each occurrence in every entity. Since there are choices of identifiers for many entities, the logical model indicates the current selection of identity. Propagation of identifiers as foreign keys may be explicit or implied.

The conversion from the logical to physical data architecture is a nonlinear process and will require additional planning and coordination between the developer and the users. Several volumetric parameters must be evaluated including:

- User population data access characteristics
- Table usage

- Table growth
- Table row populations

These items will allow the proper sizing of hardware, and, more importantly, to understand the activity of the data. Once the volumetrics are in place the next component of the architecture are the schemas and interfaces. The schemas will be created from three major components:

- The logical view of the data
- Use of the data
- De-normalization to adjust performance

De-normalization mapping will outline each of the de-normalization steps taken to improve performance.

The developer will construct the physical data model. The physical data model will provide more detail about the actual data fields than the logical data model. The documentation requirement is to have a complete data model designed using a database design tool. The database design tool should allow a person to evaluate the design using a variety of views and report formats.

Detailed System Design Specification Documentation

The developer will provide a complete set of product specifications for the system. The specifications will detail all aspects of the project and will be used by the development team to code the system. The specifications should contain at a minimum the following topics:

Functional Decomposition

The objective of the Functional Decomposition is to decompose the system step by step, beginning with the main function of the system and continuing with the interim levels down to the level of elementary functions. On each level abstractions are made from each corresponding lower level. All the sub-functions together form the completely decomposed function (functional hierarchy).

Logical Application Design

Definition of modules and functional requirements of each module.

Program Interface Architecture

Form mock-ups will be designed for each of the entry screens. Each form will be accompanied by user form interaction, business rules, and input tables.

Global Features and Integration

A complete discussion of the global features that will be included in the system. These features include items such as the use of Wizards, Printing, Online Help, etc.

System Security Model

The complete system security model is defined.

Standards, Methods, Tools and Errors

Coding standards will be discussed in detail. The methods of development including version control, change control and defect tracking systems will be identified. Tools including development and reporting tools will be identified. Error checking will serve to check completeness and formatting. Error handling will serve to report internal errors to the user. Error checking will be included in the data entry and data file selection actions.

System Administration

The administration component of the system will be described including: System Administration, Site Administration and User Administration.

Calculations, Reports and Graphs

The output of the system including the reports, alerts, etc., will be described in detail.

Project Plan, QA Plan and System Test Plan Documentation

Project Plan

The project plan is a detailed proposal describing each of the objectives of the project. The project plan includes deliverable requirements and the necessary time frame for each component of the system. The Project Plan will contain a timeline for each phase of the project that identifies specific milestones including project meetings, status report submissions, and project deliverables.

Quality Assurance Plan

The purpose of this Software Quality Assurance Plan (QAP) is to define the techniques, procedures and methodologies that will be used by the developer to assure timely delivery of the software that meets specified requirements within project resources.

System Test Plan

The Test Team assigned to the project will be responsible for developing a Test Plan for each phase of the project. The contents of the Test Plan should be based on IEEE Std. 1012-1986, IEEE Std. 829-1983, and IEEE Std. 1008-1987. The Test Plan identifies the hardware, software, test personnel, location and strategy employed to test the system.

Summary

The design of a system to support processes that are governed by the GLPs must be well documented. The documentation of the system design forms the foundation of the development phase of the system. Only a well-designed system will provide the users with a high degree of assurance that the system will consistently function according to the written documentation. Therefore, when purchasing or evaluating a GLP based system, it is important to meet with the developer and understand the steps used in designing the system. The specifications are the primary document that defines the functionality of the system. All validation testing should be designed to insure the system operates as stated in the design documentation.

Chapter 16

Computer Validation in a Regulatory Environment

Ann M. Speaker[1] and Sharon M. McKilligin[2]

[1]Covance Clinical Research Unit, 309 West Washington, Madison, WI 53703
[2]Covance Laboratories, Inc., 3301 Kinsman Boulevard, Madison, WI 53704

The US Food and Drug Administration (FDA) defines Validation as follows: "Establishing documented evidence which provides a high degree of assurance that a specific process will consistently produce a product meeting its predetermined specifications and quality attributes *(1)*.

Thus, computer system/software "validation" may be regarded as a series of quality management steps taken to determine whether computer systems, either developed 'in-house' or supplied by a vendor, are able to meet the demands placed on them in terms of functionality, and that they will continue to operate consistently and reliably in a production environment.

Computer systems requiring **formal** validation are divided into two broad categories, 'compliance-critical' and 'business-critical':

- **Compliance critical** systems are those which generate or manipulate data incorporated in Clinical Study Reports, Case Report Forms, or any other form of 'raw' data. Such systems must be validated in order to fulfill the regulatory requirement for "Clinical trial data (which) are credible" *(2)*. Failure to perform adequate validation may result in regulatory non-

acceptance and lack of confidence in data generated. In a regulatory setting, effective validation ensures integrity of data from source to report/submission.
- **Business-critical** systems are those which are essential for the continued smooth operation of a clinical research unit, and could include databases, network operating systems and/or specialized software applications which support business activities.

Computer systems that are considered not to require formal validation should be subject to some User Acceptance Testing. The User should test the functionality of the system to ensure it complies with the needs of the User and the User Department. Acceptability of the system should be documented and the system released as for formally validated systems.

The benefits of software validation include, but are not limited to, assuring product quality for software automated operations, lessening risk to users, decreasing failure rates and reducing liability. Software validation can increase the usability and reliability of a system.

Software validation includes all of the verification and testing activities conducted throughout the Software Development Life Cycle (SDLC). Proper validation of software includes the planning, execution, analysis and documentation of appropriate validation activities and tasks. It should begin when design planning and design input begin and continue until the software product is no longer used.

Activities in a typical software life cycle include:

- Management
- Requirements
- Design
- Implementation
- Integration and Test
- Installation
- Operation and Support
- Maintenance

There are several distinct phases of the SDLC which will be explained in more detail; however, the basic foundation upon which the entire validation process is based is the existence of *pre-determined* and *documented* software requirement specifications. The correctness and completeness of the system requirements should be addressed as part of the process, demonstrating in the end that all of the software requirements have been met, and that all software requirements are traceable to the system requirements.

Initiation Phase

The initiation phase of a project begins when a request is made for a new automated function, or a software change is needed to correct a problem, or to enhance a software function. The request must be documented by the end user and submitted to the Information Technologies (IT) Department. New software development or purchase, enhancements or major defect corrections are initiated through the use of a Software Service Request (SSR). The SSR should identify all basic functions required by the user (e.g. primary inputs, calculations and reports). The SSR initiates the tracking process and is used to monitor the phases of the SDLC.

A risk analysis is prepared to identify the regulatory impact (e.g., data integrity, security), and an assessment of the business risk (e.g., reliance on the system, protection of assets) will be conducted and documented. This analysis shall take into account:

- Vendor history
- Purpose of the application
- Amount of data collected by the system
- Affect on current procedures, systems and applications
- Cross departmental use

The SSR is the initial formal request for software development or software modification, which initiates the concept phase. Management then weighs the business risks and decides whether or not to authorize development or procurement.

Requirements Phase

The end-user requirements are documented in a Software Requirements Specification (SRS). This document should, when applicable contain the following:

- An introduction section outlining the purpose, scope, definitions, acronyms and abbreviations, references and an overview
- A general description of the product perspective and product functions including but not limited to:

    Regulatory Policies
    Hardware Limitations
    Interfaces to other Applications
    Parallel Operation
    Audit Function
    Control Functions
    Higher-Order Language Requirements
    Communication Protocols
    Criticality of the application
    Safety and Security Considerations

- All inputs and outputs of the system
- Performance requirements(e.g. data throughput, reliability, timing etc.)
- What constitutes an error and how errors should be handled

The SRS should be reviewed for adequacy, technical feasibility and completeness, then submitted to the Quality Assurance Unit (QAU) for review.

A project management plan is developed in the Initiation Phase and should identify the specific tasks involved in the project and the timeline for completion.

Design Phase

In the Design Phase, software requirements must be translated into a logical and physical representation of the software to be implemented. The SRS is used by the programming staff to develop a Software Design Description (SDD) and, where applicable, a series of prototypes. The SDD describes data structures, information flow, control logic, parameters to be measured or recorded, error and alarm measures, security measures and predetermined criteria for acceptance. Adherence to internal programming standards ensures consistency, and technical adequacy is evaluated following formal interface, source code and database review.

At the end of design activity, a formal design review should be conducted to verify that the design is correct, consistent, complete, accurate and testable.

Development Phase

The next phase is the Development Phase, where detailed design specifications are implemented as source code. Code comments should provide useful and descriptive information for a module, including expected inputs and outputs, variables referenced, expected data types and operations to be performed. Source code should be evaluated to verify its compliance with the corresponding detailed design specifications. Source code evaluations are often implemented as code inspections and code walkthroughs. Appropriate documentation of the performance of source code evaluations should be maintained as part of the validation information.

Documentation is crucial in the Development Phase and includes all end-user manuals, unit testing summary report, a user acceptance test plan, results of the database design, results of the source code review, and a traceability analysis.

A source code traceability analysis is used to verify that all code is linked to established specifications and established test procedures. A source code traceability analysis should be conducted and documented to:

- Verify that each element of the software design specification has been implemented in code;
- Verify that modules and functions implemented in code can be traced back to an element in the software design specification;
- Verify that tests for modules and functions can be traced back to an element in the software design specification; and
- Verify that tests for modules and functions can be traced back to source code for the same modules and functions

Test Phase

The next and very critical phase is the Test Phase. Software testing objectives include demonstration of compliance with all software specifications, and production of evidence which provides confidence that defects which may lead to errors or problems have all been identified and removed.

A software testing strategy designed to find software *defects* will produce far different results than a strategy designed to prove that the software works. A complete software testing program uses both strategies to accomplish these objectives.

Test plans are created during the prior Development phase and should include a description of all tests to be run, the purpose of each test, the data sets to be used, identification of each input, and the expected output. The items included in the testing should provide a thorough method for evaluation of the following elements:

- System security
- Data integrity (storage, retrieval, audit trail)
- Measurement accuracy and reproducibility
- Calculation accuracy and reproducibility
- Stress testing by identifying factors that may cause a system failure (e.g., boundary limits, negative values, inappropriate characters)
- Completeness and utility of reporting formats
- Traceability

Test plans should identify the necessary levels and extent of testing, as well as clear, pre-determined acceptance criteria. The test plan should also include detailed instructions for testing, environment, data to be used and specific criteria for acceptance. Test results should be documented to allow objective pass/fail decisions to be reached. Errors detected during testing should be logged, classified, reviewed and resolved prior to release of the software.

The testing must be conducted within a simulated production environment or environment identical to that which will be used in production. A test log is used to record the actual results of the testing, then a test report is generated to present the results of tests performed and described in the plan. It will present conclusions regarding the success of each test based on the criteria specified.

The release of software for testing and use in a production environment will be done under configuration management by the system administrator. Configuration management ensures strict version control of the software.

Installation Phase

Prior to placing a system in production, the user acceptance test plan and test report must be completed, reviewed, signed off and archived. Corresponding SOP and user manuals have been completed, staff has been trained in the use of the software and training is documented. a change control log has been created and will be maintained. A qualification plan has been written and appropriate workstations will be qualified prior to placing the software into production.

When all users have been trained and notified of the production release date, IT staff will place the system into production.

Operations Phase

All modifications, enhancements or additions to existing software or its operating environment are design changes and are subject to design control provisions. The validation activities associated with each software change should be documented as part of the record of that change. All ongoing changes must be tracked for administration and control of validated systems. These types of documentation include error logs, change control logs, qualification logs and activity logs.

The purpose of the Software Development Life Cycle is to maintain software in a validated state.

References

1. FDA Glossary, 1995.
2. ICH Good Clinical Practices, Section 1, 1997.

Chapter 17

Electronic Data Archiving: Ensuring Accessibility, Durability, and Usability

Edward J. McDevitt

DuPont Crop Protection, DuPont, Wilmington, DE 19880–0038

> The amount of scientific data generated in current times is growing at an ever-accelerating rate. The desire, need, and requirement to collect and maintain these data in a readily accessible and tamper-proof way that ensures a high degree of integrity over an indeterminate number of years exist. Changing at an equally accelerated rate is the technology used to collect, store, and retrieve data. This chapter describes the challenges for long-term data storage. It profiles the various strategies necessary for maintaining an electronic data archive.

Introduction

History is full of examples of human beings trying to preserve data, information and knowledge for future generations. The ancient libraries at Alexandria, Dead Sea scrolls, oral story telling, paintings on cave walls, stain glass windows of the Middle Ages, monastic scriptoriums, and the National Archives of governments around the world are evidence of this need. Each of these examples is different enough to demonstrate the problems inherent in the method of archiving used, be it languages no longer spoken, transcription errors, media that are fragile, media that are not portable or sheer volume. These examples also demonstrate that archiving information cannot be a one-time event for a given set of data, but a process that needs to be managed for the length of time the data, information and knowledge are believed to have value. Failure to set up a process will result in lost data, information and knowledge.

Similar issues exist in the electronic age. It would be hard to argue that the Information Technology world has not had a positive impact on science and the business world. The ease of electronic data creation and collection has opened new ways to model and solve complex problems to a level of precision never before imagined. The breadth and depth of Information Technology capabilities continues to grow and expand. It is pervasive in most things we do. It is not, however, without challenges to the preservation and the accessibility of data, information and knowledge for future generations. Software and hardware become obsolete. New products and new versions of existing products are released regularly. People and organizations rush to embrace the promise of new functionality, ease of use and performance, often giving little thought to the data stored in the current systems. The persistent reality is that technology changes will always be with us and that a process to manage this change is necessary for the successful preservation of data, information and knowledge for future generations, and for maintaining a high degree of integrity of such data, and where necessary for legal defensibility.

John Carlin, Archivist of the United States, summarizes, "Electronic records pose the biggest challenge ever to record keeping in the Federal government and elsewhere. There is no option to finding answers... the alternative is irretrievable information, unverifiable documentation, diminished accountability, and lost history."

Definitions

Basic terms for records management are defined in 36 CFR 1220.14. The following are found in 36 CFR 1234 --

- Database - a set of data, consisting of at least one data file that is sufficient for a given purpose.
- Data base management system - a software system used to access and retrieve data stored in a database.
- Data file - related numeric, textual, or graphic information that is organized in a strictly prescribed form and format.
- Electronic information system - a system that contains and provides access to computerized records and other information.
- Electronic record - any information that is recorded in a form that only a computer can process and that satisfies the definition of a Federal record in 44 U.S.C. 3301.
- Electronic record keeping system - an electronic system in which records are collected, organized, and categorized to facilitate their preservation, retrieval, use, and disposition.

- Text documents - narrative or tabular documents, such as letters, memoranda, and reports, in loosely prescribed form and format.

Strategic Components

Three strategic components of a successful electronic data archiving process are:

- a strong Records Management program
- a mature Information Technology (IT) Life Cycle Management program
- an organization committed to the principles of these programs

All these components must work in concert with one another.

To be successful, an electronic data archiving process needs a well-developed Records Management program that defines rules by which records and documents are governed from creation to disposition. Records are classified by type such as research records, personnel records, tax and financial records, etc. Each record type has a defined retention time.

The preamble for DuPont's Corporate Records and Information Management says:

"Proper records management is an important function of every successful corporation. An effective records management program ensures that all records that are required to conduct the business of the corporation, to fulfill its legal responsibilities, and to support its tax liabilities are maintained and available.

"An effective Records Management program also preserves the corporate memory and protects the corporation by ensuring compliance with local and federal laws.

"Significant costs are associated with the creation, maintenance, distribution, and storage of records. Therefore stewardship must be exercised..." - DuPont Corporate Records and Information Management preamble.

To be successful, an electronic data archiving process must have a well-developed IT Life Cycle Management program that defines the rules governing the four life cycle phases: introducing new technology, mainstreaming technology, containing technology and retiring technology.

Retiring technology requires decisions as to what data and functionality continue to migrate forward based on the rules set out in the Records Management program. It is often a handoff to technology in the introductory phase. This process presents an opportunity to reassess the enduring nature of the data stored in the retiring technology.

Laboratory technology, for example analytical equipment, not traditionally thought of as a component of Information Technology, needs to be a defined part of IT Life Cycle Management for R&D and manufacturing for those devices that generate, store, transmit or render data. Analytical equipment, today, is fully IT enabled. They have PC controllers, processors, and disk drives. They have access to the network, on board software for collection, reduction and rendering.

Generally speaking, IT Life Cycle planning today is often project, reaction, or necessity-based rather than based on a well-maintained master plan. This is not necessarily bad. Projects are sponsored by the local organization and hence project teams are closer to where the needs and recordkeeping rules are defined. However, the project teams need to understand the technology directions of the larger organization. This ensures that the proper infrastructure is in place to support the production system.

The following are examples of technology changes in industry: Macintosh computers were the tools of choice for many years in R&D environments. Industry convergence and corporate policies moved many organizations to Window PCs. Similarly, email systems changed. Some moved from All-in-one to LotusNotes or Microsoft Outlook. Relational databases changed with many choosing Oracle or Microsoft Access. File and Print Servers also changed with many organizations moving from Novell to Windows NT.

Each of these technology transitions offers a different set of capabilities and limitations. For example, fonts available in Macintosh Microsoft Office may not be available in Microsoft Office for Windows. In addition, software may be available on the Macintosh but may not be available on the Windows platform. Each of these technology changes required careful planning and project management to ensure no disruption to the organization and no loss of data.

There is also a tactical component of IT Life Cycle Management, which is physical media management. Tapes, disks, and other electronic media degrade over time. They need to be refreshed every ten years. Ideally, this is part of the standard operating procedures for the data centers and archive facilities.

Finally, an electronic data archiving process can only be successful with a committed organization with ethical individuals supporting Records Management and IT Life Cycle Management programs. Non-IT members of the organization need to assume stewardship roles over the data, information and knowledge generated by their organizations. Senior management must support the enforcement of these programs as well as understand the need to track evolving government regulations in the area of record management. After all, electronic data, information and knowledge are valuable organizational assets.

Managing Accessibility

Accessibility is defined as the ability to locate and use data, information and knowledge known to exist in the organization. In the ideal world there would be a master index containing pointers to all data and information stores in the organization. This is typically an unrealistic expectation. The "means and will" required for maintaining such a repository is high. Most organizations manage the "master index" within small groups with varying degrees of formality. Documents and data required for regulatory or patent purposes are often given very formal attention. Different organizational requirements will dictate the rules on accessibility.

Availability is an aspect of accessibility. The organization may need anytime and anywhere access. Speed of access to historic data is sometimes a requirement. Recall times can vary. They may be hours, days, weeks, etc. The requirements will most likely depend on the type of data being requested. There are also certain cost implications depending on the requirements needed. On-line storage is most convenient but as repositories of data grow, system performance maybe effected. Hardware and software must be scalable to accommodate such potential growth. Data centers typically charge a premium for such ready access. Off-line storage on tapes or CDs is often less costly but carries with it the latency of having to retrieve and load the data set. Off-line may also mean off site.

Fundamentally, however, those records, that are deemed to have enduring value and need to be preserved and available, must be identified and planned for in advance, instead of later reacting to technological change, which could jeopardize access.

Disaster recovery plans and appropriate levels of system redundancy also help ensure access to data, information and knowledge.

Managing Durability

Storage media ages. It is necessary to refresh the physical media about every ten years. Storage media options also change. Eight-inch and five and a quarter inch diskettes are hard to come by. Similarly, magnetic tapes and tape drives change requiring the transfer of stored information to new media types. One way to address this problem is to maintain outdated equipment. This, however, simply delays the inevitable need to migrate. Maintaining old storage equipment is arguably just as expensive as migrating to new media and new equipment. Parts and service become scarce and expensive. The procedures used to transfer from media to media or media type to new media type must be validated and QA'd to ensure accuracy and reliability in the new copy. Backup and recovery procedures may also need to change. The Records Management principles require the destruction of the old copy once the new copy has been successfully created.

What is true of the storage media is also true of the software used to store and access the data. Software versions become out of date and are no longer vendor supported. People with the skill sets necessary to support the software are expensive and become difficult to find. The need to migrate to new versions is a necessity. The new software version and the migration plan need to be validated and QA'd from a data preservation and functional need perspective.

In May 1996, the US Task Force on Archiving Digital Information reported that:

"Neither 'refreshing' nor emulation sufficiently describes the full range of options needed and available for digital preservation. Instead, a better and more general concept to describe these options is migration.

"Migration is a set of organized tasks designed to achieve periodic transfer of digital materials from one hardware/software configuration to another, or from one generation of computer technology to a subsequent generation. The purpose of migration is to retain the ability to display, retrieve, manipulate and use digital information in the face of constantly changing technology. Migration includes refreshing as a means of digital preservation but differs from it in the sense that it is not always possible to make an exact digital copy or replica of a database or other information object as software and hardware change and still maintain the compatibility of the object with a new generation of technology."

Managing Usability

It is not enough to migrate only "raw" data – the characters, the numbers, the bit and bytes – forward to ensure usability. The metadata and the context for the application or database must also be migrated forward. Metadata is the code to the machine-stored bits and bytes. Metadata is the data about the data. It describes the data in the database.

For example, it indicates that a field or column called LAST_NAME exists and is 40 characters wide. The metadata indicates this is a secondary key to a table called TEST. This is a required field. The additional system and user documentation further indicates that this is the last name of the experimenter performing the procedure.

The metadata documentation describes the method of data capture, the application used to access the data, security rules for the tables and columns as well as other descriptive and procedural information. For derived or calculated data, it is important to know what algorithm or protocol was used. The documentation then becomes something else that needs to be preserved.

It is important to note that without the metadata in the above example the reader will only see a series of alphabetic characters. Without the entire described context associated with the data, it has no meaning.

An example of an issue that can arise when the old metadata does not cleanly carry forward is the use of a DATE field in an older research database. These data were generated in the 1930's. They were first recorded electronically in the 1960's in a system that did not have a modern date field. In this example, the date was simply recorded as month and year. This met the organization's needs and technical capabilities at the time. In the new database, the date is the full date of day, month and four-digit year. Another example is a NAME field in an older system that in the migrated future version is recorded as FIRST_NAME, MIDDLE_NAME, and LAST_NAME.

Conversion rules need to be defined and documented. This involves the IT organization and the organization's stewards for the data being converted.

Additionally, if there are data quality problems, it is important to address them prior to archiving and migrating. The act of archiving, by itself, will not improve data quality. When the data are retrieved at some future time, it will be difficult, if not impossible, to address and correct data quality issues. If the data, information and knowledge are said to have enduring value, then the quality must be kept high throughout their retention period.

Conclusion

The need to develop an electronic recordkeeping strategy is critical and fundamental to the success of retrieving archived data, information and knowledge that has enduring value for future generations. This strategy must be developed within the current Records Management process in partnership with both the Information Technology Life Cycle Management program and the stewards of the data, information and knowledge. The latter are the non-IT managers in the organization

Value determination is not the job of Information Technology; it is the job of the data, information and knowledge stewards. These managers are responsible for determining the rules by which the data will be selected, archived and retrieved. It is the job of Information Technology to maintain and execute these rules using the appropriate technologies over time.

It is important to set expectations on the archiving process. The organization needs to know what can actually be brought back and in what form. It needs to know to what level of completeness the data can be retrieved. It also needs to know how much confidence it can place in the retrieved data.

Wherever possible, in the Information Technology selection process, establish the need for common usage rules among systems, compliance with record keeping standards, and technology that supports record management and migration.

Acknowledgements

I would like to thank my colleagues at DuPont for their help and input in bringing this paper together, in particular Iris Fisher, Manager Corporate Records Management team, Aster Wu, Data Architect, Cecelia Smith, Information Management, Sandra Hughes, Application Specialist and Bruce Lockett, Analytical Sciences. I would also like to thank the US National Archives and the Australian National Archives for their availability by telephone and for the wealth of information made available to the public.

Bibliography

1. Australian Government, "A Guide to the Metadata Fields of the Marine and Costal Data Directory for Australia, Blue Pages", September, 1997
2. Kingma, Bruce, "The Cost of Print, Fiche, and Digital Access", D-Lib Magazine, February, 2000
3. Moore, Baru, Rajasekar, Ludaescher, Marciano, Wan, Schroeder, and Gupta, "Collection-Based Persistent Digital Archives – Part 1", D-Lib Magazine, March, 2000
4. Moore, Baru, Rajasekar, Ludaescher, Marciano, Wan, Schroeder, and Gupta, "Collection-Based Persistent Digital Archives – Part 2", D-Lib Magazine, April, 2000
5. National Archives of Australia, "Keeping Electronic Records", http://www.naa.gov.au/recordkeeping/er/keeping_er, March 1995
6. United States National Archives and Records Administration, "Electronic Records Management", 36 CFR Part 1234, last amended July, 1998
7. United States National Archives and Records Administration, "Fast Track Guidance Development Project", January, 1999
8. United States National Archives and Records Administration, "Transfer of Electronic Records", 36 CFR 1228.270
9. Waters, Garrett, "Preserving Digital Information", Commission on Preservation and Access, May, 1996

Products Mentioned

1. Microsoft Access, Microsoft Office, Windows, Windows NT are products and registered trademarks of Microsoft Corporation

2. Oracle RDBMS is a product and registered trademark of Oracle Corporation
3. Macintosh is a product and registered trademark of Apple Computer Inc.
4. LotusNotes is a product and registered trademark of Lotus Development Corporation
5. All-in-1 is a product and registered trademark of Compaq Computer Corporation
6. Novell is a product and registered trademark of Novell Inc.

Chapter 18

A New Order of Things: Electronic Pesticide Submissions: The Promise Is Efficiency

William N. Casey

AVENTIS CropScience, P.O. Box 12014, Research Triangle Park, NC 27709

The key to making a successful electronic submission will depend on the ability of the industry to publish documents electronically and on paper. The electronic version must be identical to the paper copy while offering the ease of use of the WEB.

The phenomenal growth of the INTERNET has revolutionized the delivery of text and image based information. All the signs are that this will be the definitive technology for the next five years and maybe more. The rate of change in computing will pull us all forward and we may not be in the position to drive such changes but merely to follow. If you look at the pharmaceutical sector you will find that "Electronic Submissions" are already being accepted by the FDA. In the pesticide area we are currently the followers.

Where does one begin? You hear acronyms like CADDY, PDF, HTML and XML, but what exactly would an electronic submission look like? How would it function? What is the basic building block of an "Electronic Submission?"

The success of electronic submission will depend on the ability of the industry to publish documents electronically and on paper. The electronic version must be identical to the paper copy while offering the ease of use of the WEB. Scanning studies is not an option.

Our current way of thinking is in the direction of "How does one get one's data onto paper?" Perhaps a better way to think is "How does one get one's data into an electronic document that will have the format and appearance of paper but with a structured content?" In order to accomplish this, it will call for a significant change in the work habits of those who write reports. We must therefore initiate a new order of things. *"There is nothing more difficult to carry out, nor more doubtful of success, nor more dangerous to handle than to initiate a new order of things."* - Machiavelli, The Prince, 1532

When many of us began our careers, reports were prepared for publication using typewriters and a support staff of typists. The first change that introduced electronic capabilities for document production occurred with the introduction of word processing systems such as WANG. While this represented progress, the greatest efficiency gain did not occur until the introduction of the personal computer. With the introduction of the PC into the work place, we saw the first significant change in how reports were written. Complete control of the publication of a report now can lay solely in the hands of its author. Though some Luddites were reluctant to adopt this new technology, today a professional preparing a report on a PC is common place and almost taken for granted. New skills were required but the gain in efficiency and control over one's document led to a rapid acceptance of the PC and a shift in how businesses organized document production.

While the PC offered more control and efficiency, a complete report including appendices was not always possible to produce in a word processing file. Also, the ability to share with others was limited due to compatibility problems from PC to PC. Even when using the same software format shifts and printing problems could occur. In other words, what appeared on the screen was not always the same as the printed copy. As a solution, the scanning of documents and the storing of each page as an unalterable image offered for the first time the ability to prepare a pesticide dossier electronically. While this method offers a reduction in the volume of paper, there is no real efficiency gain to be had with this strategy for regulatory reviewers. Also, there is no real change in how documents are produced; the thought process is still "paper first, electronic version second."

Just what do reviewers want? A good example can be found in FDA's guidance to industry on electronic submissions. In their guidelines FDA indicated that documents submitted in electronic format should:

- Enable the user to easily view a clear and legible copy of the information.
- Enable the user to print each document page by page, as it would have been provided on paper, maintaining fonts, special orientations, table formats, and page numbers.
- Include a well-structured table of contents and allow the user to navigate easily through the submission.
- Allow the user to search and copy text and images electronically into common word processing documents.

To achieve the above goals, FDA has required that electronic documents be submitted in Portable Document Format (PDF). The pesticide regulatory agencies in North America (EPA and PRMA) have followed FDA's lead. PDF's property of capturing the exact intent of a document's design in a "final form" is critical to providing regulatory agencies with documents that can transform how one currently works with pesticide submissions. PDF enables document submitters to provide things like annotations, table of contents, bookmarks, hyperlinks within a single document and between other documents, links to supplemental files in their original form, and content searching. Critical is the ability to cut and paste information for reuse and further analysis. But the most important point is the ease of producing PDF documents from any application that has the ability to print, and to have as a result, an electronic version which is identical to the paper version. Both form and structure are needed in electronic pesticide dossiers and the report or study level is the initial building block. With studies in a PDF format, a dynamic submission dossier can be built containing all of the features needed to reduce effort and gain efficiencies within both the regulatory agencies and industry.

Creation of PDF Documents

Where does one begin and what are the tools required to publish a study in PDF format? Actually the process can be relatively simple from the standpoint of a single study. The minimal tools required are: Adobe Acrobat 4.0 and a plug-in, such as, Impress from Mapsoft Computer Services Ltd. to provide document pagination. Acrobat offers several methods for converting an electronic file from virtually any application to the Portable Document Format (PDF). One can use either one of two utilities to make this conversion: PDFWriter and Acrobat Distiller. PDFWriter is a printer driver that converts files directly to PDF from any another software application. Distiller is a tool that converts PostScript files to PDF, and provides higher quality output than PDFWriter. PDFWriter is often quicker to use than Distiller. Distiller maintains all the formatting, graphics, and photographic images from the original document, and it provides more precise control over the conversion process than PDFWriter and would be the preferred option for complex documents. Creating PDF files using Microsoft Office software is as easy as printing and bookmarks can be created automatically based on standard MS Word headings. Macros included in the Acrobat installation place a "Create Adobe PDF" as an option under the "File/Print" menu of MS Word and MS Excel so that one can create PDF documents directly from these applications. The basic steps in the creation of a report in PDF file format are as follows:

- File conversion from applications (i.e., MS WORD MS Excel, and others) containing report components by using PDFWriter or Distiller.
- Assembly of PDF files into final report.
- Scan signature page(s) and insert into final report.
- Final document and table of contents pagination.
- Set additional bookmarks if needed (table of contents) to provide navigation.
- Add document reference numbers (e.g., unique report number or EPA MRID number.)
- Print the final paper copy from the PDF file.

Just as one can print sections or parts of a report from different applications and assemble the report by inserting the paper pages in its appropriate place, one can also insert pages in PDF format from one PDF into another PDF. It is "cut and paste" at the page level. PDF can capture the exact intent of a document's design in a "final form" that is identical to the paper version. This is critical to the success of electronic pesticide submissions. The ease (simplicity) of creating a PDF from any application with the ease of printing is a very powerful feature. The construction of the different components of a study needs to be considered based on page(s) insertion and deletion capabilities of PDF. For example, different sections of a report can originate from different applications; tables from Excel can be combined with text sections from Word by rendering them to separate PDF files and then assembling them together into a single PDF.

PDF is a file format that represents a document in a manner independent of the hardware, operating system, and the application software used to create the document. PDF was developed to allow documents to be transferred and shared across computer platforms. This capability allows one to construct a single document under one common format from many different applications.Benefits of PDF Documents

Benefits of PDF Documents

What are the advantages to the reviewer? Simply put, an increase in efficiency. A time savings can be obtained for reviewers in preparing data evaluation records due to the ability to copy and paste (drop and drag) information from the PDF study directly into the reviewer's application.

Acrobat 4.0 has a table/formatted text select tool that allows one to select tables and text in a PDF document and retain the original formatting when the material is copied (or imported) into other applications. One can specify vertical or horizontal format, the type of text flow, and whether one wants ANSI (simple text) or Rich Text Format (RTF).

- Select the table/formatted text select tool. The cursor then changes to a crosshair.
- Drag a rectangle to enclose the table or formatted text you want to copy.
- Press the right mouse button, and choose the necessary options.
- Drag the selected table or text to the desired application.

Being able to search the dossier for relevant information without having to manually search through several studies page by page is another benefit. Imagine being able to search a chronic toxicology study and track a single animal's reference by being able to see just the pages where the reference occurs. This is a big time savings over reviewing each page of a 2000 page report. The indexing of an entire pesticide submission is possible when using searchable PDF files. This allows full content searching across all documents in a collection. For example, one may search for a metabolite and find where it is referenced on all pages in each study of a submission.

What is industry looking for as a benefit? First and most important is a reduction in time to market. Increased efficiency in report production. Longer term, libraries of studies that permit full content searching (knowledge management) reducing the time to respond to questions. Documents entering an electronic document management system would no longer have to be scanned. The same benefits for the reviewer also apply to industry scientists in being able to reuse data into a new document by being able to copy and paste formatted data regardless of platform and the application used to create the original study.

Conclusion

Electronic study production will become a standard practice in the Agro Chemicals industry within the next 3 years. Electronic pesticide submissions are possible provided studies are constructed from the ground up based on electronic files or components. The ability to construct a complete electronic study in a "final form" identical to the paper copy is provided by PDF today. Be ready for change. New skills must be learned. IT organizations must provide support for the document production process. And finally, a new order of things, think PDF first; paper last.

Chapter 19

Benefits of Electronic Data Submissions: An Industry Perspective

Thomas J. Gilding

American Crop Protection Association, Suite 400, 1156 15th Street, N. W., Washington, DC 20005

The interest in electronic data submission (EDS) for the pesticide registration process has gained significant interest among government regulatory authorities and registrants. The technology has evolved to the stage where valuable efficiency gains are more evident and closer to becoming reality. However, to successfully achieve the benefits of EDS, the applications where the benefits are the greatest need to be clearly identified. The American Crop Protection Association (ACPA) believes that the greatest benefits from EDS lie within the functions of government reviewers of registration applications. The efficiency gains from EDS in the government review process should translate to shorter application review times for registrants. A government/industry partnership is critical to implementing EDS into the pesticide registration process. The primary purpose for this partnership is to ensure consistency in data submission and review requirements, and compatibility of computer hardware and software. Pilot submissions by registrants should be used to validate specific electronic applications and verify potential efficiency gains. This chapter reviews the past, present and future perspectives of ACPA on approaches for implementing EDS into the U.S. pesticide registration process and internationally. ACPA is a non-profit trade association representing companies who manufacture, formulate, distribute and sell pesticides for U.S. agricultural production.

© 2002 American Chemical Society

The potential benefits that electronic data submission (EDS) offers to the pesticide registration process have been recognized to a limited extent over the past ten to fifteen years. This primarily existed in individuals who were involved in the laborious tasks of managing pesticide registration data and had some knowledge in computer systems. Just knowing the ease and speed of handling data electronically and looking at the extensive amounts of data and mountains of paper required in making pesticide registration applications, made it quite obvious to them that EDS could only make things better.

During this period, a number of projects were initiated by EPA, ACPA or jointly, each one addressing preliminary attempts of submitting data electronically into the pesticide registration process. The projects ranged from simple submission of text files to more complex formats for study and raw data submissions. The purpose, progress, and problems of these projects provide valuable lessons for the current, more comprehensive efforts to make EDS benefits a reality.

Awareness of the potential benefits that EDS presents has increased substantially in recent years to where those in the registration process who would be the primary benefactors are becoming more involved. This includes company scientists/registration managers and EPA risk assessors/managers who are showing interest in achieving efficiency gains from EDS in the submission, storage, review, and archiving of pesticide regulatory information in contrast to current paper submissions. The principal goal for implementing EDS is to reduce the time and effort required for registrants to prepare and submit pesticide applications and for the regulatory authorities to review for reaching final registration decisions. Therefore, the definition for EDS efficiency gains means time and resource savings to both registrants and EPA reviewers in their respective functions in the registration process. ACPA translates these savings to a primary goal of achieving reduced times from submission of a registration petition to a final registration decision. The latter goal is extremely important to registrants because time is money in relation to when a product can be marketed.

The major focus for making EDS a viable efficiency tool in the pesticide registration process is to identify those functions where benefits are the greatest. ACPA believes that the greatest benefits for registrants lie not with their own functions, but in improving efficiencies for EPA reviewers. Improving the performance of EPA reviewers contributes most for achieving shorter review times. Efficiency gains in the transmittal and archiving of data submissions are important, but secondary.

The successful implementation of EDS in the pesticide registration process demands that EPA and registrants work together in a cooperative partnership to: 1) Identify the primary needs of reviewers and where EDS offers the greatest benefits; 2) Find the technology that best supports these needs; 3) Ensure that EPA's and registrants' computer hardware/software, operating procedures, and infrastructure are compatible; and 4) Verify efficiency gains and operational compatibility through pilot submission projects.

Historical Perspective

Over the past ten to fifteen years, there has been a number of small development projects on specific EDS applications in the pesticide registration process. Although narrowly focused and not broadly known as to their outcome, they are without a doubt significant in representing a start of a process that is finally gaining momentum and is what brings us to this point today. The experiences gained from these projects need to be identified and evaluated for utility in defining EDS applications that facilitate the organizing, transmitting, reviewing and archiving of pesticide data and information. Some of these projects included:

- **Report Formats:** In early 1990s, this project, initiated within EPA, examined methods for submitting study reports on computer diskettes. Several registrants participated in pilot demonstrations. The final results from this project are not known.
- **Data Evaluation Records (DERs):** In 1995-96, this was a joint ACPA and EPA project to develop formats for Data Evaluation Records (DERs) that would allow registrants to insert summary information from studies into these formats. The objective for the formats was to facilitate EPA reviewers' ability to review, evaluate, provide supplemental information, or revise as appropriate. This project focused on toxicology studies, but was not completed because of higher priorities and resource demands created by FQPA.
- **Standardized Residue Chemistry Study Report:** In the mid-1990s, ACPA (then the National Agricultural Chemicals Association, NACA) initiated a project to standardize formats for submission of residue chemistry studies. The objective was to have residue study reports list information in the same sequence to facilitate the tasks of EPA reviewers. This project was successfully completed and the formats are in use.
- **Toxicology Data Format:** In early 1990s, the ACPA (then NACA) Toxicology Work Group on Electronic Exchange of Data (NTWEED, later ATWEED) worked with EPA to develop an electronic format for

submitting data from a particular toxicology study and obtain ANSI approval. The final results from this project are not known.
- **Electronic Information Exchange and Interaction Organization (EIEIO):** In mid-1990s, this was a joint ACPA and EPA project to explore the feasibility and practicality of registrants submitting technical reports electronically. This project was not completed because of higher priorities and resource demands created by FQPA.
- **Computer-Aided Dossier and Data Supply (CADDY):** In 1995, the European Crop Protection Association (ECPA) and the European Commission initiated a project to develop and implement a system for the electronic submissions of dossiers to the European Union for pesticide registrations. This effort resulted in the CADDY specification and supporting retrieval software. CADDY is a cost-saving, paper reduction and achiving tool. As of the end of 1999, approximately 22 CADDY submissions were made to EU countries.

These projects raise an interesting question of why EPA and registrants have taken so long in becoming serious about implementing EDS applications for the pesticide registration process. A number of reasons come to mind, e.g., the technology is still in its infancy, "finger pointing" between EPA and registrants on who should take the lead, and there are certainly others. Probably the most probable reason is due to the lack of demonstrated economic value to both EPA and registrants necessary for raising management interest and providing justification for expending resources for development, start-up and implementation. Another reason, at least for explaining the hesitation of registrants, has been their unwillingness to embark unilaterally, spending money on computer system capabilities without knowing what EPA's final requirements for EDS procedures and computer hardware/software will be.

Several major activities have started over that past several years that are "giving direction and making things happen." The Pest Management Regulatory Agency (PMRA) of Canada began a very aggressive program to implement Electronic Dossier Delivery and Evaluation, EDDE, (PMRA's term for EDS). PMRA's motivation for the program is driven primarily by a cost recovery mandate from its legislature. One of the first actions taken by PMRA was to identify the requirements of both its reviewers and registrants. It then made decisions on selecting EDS applications and supporting computer hardware/software technology to meet these requirements. PRMA is actively seeking pilot submissions to verify its technology and procedural selections.

Another important aspect of the Canadian program is in promoting systems compatibility, not only with Canadian registrants, but also other national regulatory agencies to facilitate international harmonization of registration processes. PMRA is presently in a second revision of several draft guidance documents in support of implementing its EDDE program. These documents are: 1) Guidance for Electronic Exchange; 2) Statement of Evaluator Requirements; and 3) Guidance for Industry During Pilot Stage. ACPA commented to PMRA in October 1999 on the above three draft documents.

Because PMRA and EPA are working closely together on pesticide registration harmonization under NAFTA, EPA's approach for implementing EDS is similar to that of PMRA. In addition, EPA and PMRA are working together to harmonize EDS applications through their involvement in the NAFTA Technical Working Group on Pesticides (NAFTA TWG). ACPA supports the approach being taken by EPA. EPA and ACPA have agreed in principle to coordinate our respective efforts relating to EDS implementation for domestic registrations. The first step in this coordination was a joint workshop held in the Washington, D.C., area on March 28-29, 2000.

The purpose of the workshop was to bring together EPA reviewers and registrant scientists to discuss issues, barriers and solutions, for making EDS a viable tool in the pesticide registration process. At the workshop, the 80 participants were first given an overview of EDS positions and experiences of government and industry representatives. These presentations provided a basis of orientation and set the stage for more detailed discussions in concurrent breakout sessions. This workshop was successful in starting constructive dialogue between EPA and registrants. A follow-on workshop is tentatively planned for sometime in 2001.

The Food and Drug Administration (FDA) has valuable experience in EDS applications for pharmaceutical drug registrations that should be evaluated closely for application to pesticide registrations. FDA has a legislative mandate to accept all submissions in electronic format by the year 2002. The regulations for accomplishing this can be found in the Federal Register, 21 CFR Part 11. FDA is very much on its way to meeting this legislative mandate and EPA is looking to see what relevant experience it can utilize for pesticides. ACPA plans to do similar coordination with the Pharmaceutical Research and Manufacturers of America (PhARMA) to benefit from the pharmaceutical industry's experience.

From a global perspective, several items are worth mentioning. The Global Crop Protection Federation (GCPF), a federation of national associations of the crop protection industry of which ACPA is a member, is focusing on greater coordination between North American (Canada and U.S.) and European industry approaches to EDS implementation. This primarily involves defining the needs of our respective regulatory authorities to identify common requirements on which coordinated global industry approaches can be based. GCPF is also considering proposals to the Organization for Economic

Co-operation and Development (OECD) Working Group on Pesticides to begin EDS harmonization projects similar to the ongoing efforts of the NAFTA TWG. The OECD Working Group on Pesticides membership consists of pesticide regulatory authorities from 29 countries. Obviously, achieving harmonization within the OECD countries will be much more difficult than achieving harmonization among the 3 NAFTA countries, but then again the benefits will also be much greater.

ACPA Position and Perspectives

One critical objective in ACPA's strategic planning calls for working with EPA on improving timeliness, productivity and efficiency in the pesticide registration process. We view EDS to be an important means for achieving this objective through significant efficiency gains over the present system involving paper submissions. ACPA's perception of the term efficiency gains is not only bringing cost-saving measures into the registration process, but also that the measures result in reduced times for registrants from when registration petitions are submitted to a final registration decision by EPA. Our member companies are very clear on the importance they place on time saving benefits of EDS. After all, time is money when one considers missed opportunities in marketing any commercial products.

It should be clarified here that in seeking shorter review times for pesticides, our industry is not placing economic objectives over its responsibilities for assuring that its products are safe to use. To the contrary, ACPA believes that EDS can improve the quality of EPA's risk assessments while reducing the time and resource demands for both registrants and EPA reviewers.

Therefore, ACPA's expectations for EDS is the streamlining of the time and tasks required for registrants to prepare applications for submission and for EPA risk assessors and managers to review them to the extent necessary for reaching final registration decisions. We agree that the emphasis on efficiency gains from EDS should focus on the tasks of the EPA reviewers necessary to accelerate approval times. Archiving and transmittal of submissions are viewed to be important, but secondary benefits.

The fundamental principles behind ACPA's support for EDS in the pesticide registration process are that:

- There must be recognized benefits to both government reviewers and registrants with applications tailored to those needs where efficiency gains are the greatest;

- The efficiency gains from EDS must be sufficient to justify initial and re-occurring investments by registrants for the required computer hardware and software;
- Security protection for confidential business information (CBI) must exist for any government electronic submission system;
- EDS capabilities should be developed through an industry and government partnership to assure consistency in data submission and review requirements, and compatibility of computer hardware and software;
- Accuracy and reliability of EDS systems must be assured;
- EDS systems must be based on open and flexible standards;
- Pilot submissions should be the basis for verifying that specific EDS applications are operationally viable and provide expected efficiency gains; and
- The use of EDS by registrants must be optional.

Presently, neither EPA nor registrants are in the position to utilize EDS to its full potential. In fact, the full potential of EDS applications has not really been identified or demonstrated. This fact alone defines the need, scope and objectives of EDS implementation programs. ACPA stands ready to work with EPA on behalf of U.S. registrants to make EDS a viable tool in the EPA pesticide registration process.

Future Directions

Successful implementation of EDS will take a number of years and most likely be an evolving, step-wise process. It has only been one or two years now that more favorable conditions exist where government regulatory agencies and registrants are recognizing the value of EDS and the importance of working together on its implementation. The Canadian PMRA deserves recognition for advancing this favorable environment as well as the work of the European registrants in developing CADDY with their regulatory authorities. What is important now is to build on this recent progress and further identify and commit to plans that "give direction and make things happen."

ACPA has established a work group that among other tasks is working with EPA to define an EDS implementation plan for U.S. registrations. We will also be coordinating with our industry colleagues on the global scale to fulfill registrants' role and commitment to make EDS a reality in NAFTA and OECD countries. To accomplish all this, well-defined plans need to be developed through an industry/government partnership and followed up with specific actions. ACPA believes that these action plans will need certain flexibility for responding to unique national requirements; however, their overall framework and direction must be guided by the following fundamental considerations.

Needs of Government Reviewers Must be the Primary Basis for Defining and Implementing EDS Applications

The two major factors to the success of EDS implementation are whether the applications identified present the greatest benefits and if the intended users of these applications will accept the technology. ACPA agrees that EPA reviewers will be the prime users of EDS applications; therefore, the first step to implementing EDS should be the identification of those EPA reviewing functions where EDS applications are believed to present the greatest potential for efficiency gains. Examples of the kinds of reviewer functions that should be considered:

- Navigation potentials to permit page advancing/reversing, go to page capability, and hypertext linking;
- Full-text search capabilities within reports or through an entire submission;
- Capability to print selected text, pages, or range of pages;
- Creation, management, searching and sharing of annotations;
- Capability to extract text and tabular data from the submission via copy/paste mechanisms for further analysis or re-use in review reports;
- Ability to use supplemental files for extracting larger data sets where information cannot be electronically extracted in an acceptable manner from reports;
- Capability to search for reports by attributes, such as author, source, title, etc.
- Ability to create bookmarks and search for text within bookmarks;
- Ability to clearly identify, in subsequent dossier submissions, data which has been changed from, or added to, a previously submitted version;
- Ability to easily view clear and legible copies of the information, with legible size text.

The objectives of EDS implementation projects must be to verify that selected applications do result in meaningful efficiency gains acceptable to the reviewers' mode of operations and provide compatible interfaces between EDS systems of EPA and registrants.

Pilot Submissions Critical to EDS Applications Verification

Pilot submissions are critical components to the successful implementation of EDS. They are necessary to demonstrate performance of candidate EDS applications, verify compatibility between EPA and registrants computer hardware/software systems, and provide learning curve opportunities. Although it is the decision of the registrants to participate in a pilot submission, ACPA's role should be actively involved with EPA in defining on-going needs for pilot submission projects and then promote registrants' participation. There should be tangible incentives for registrants to participate in pilot stage programs. Factors that will encourage their participation are: 1) Incentives need to be identified at the start to justify costs; 2) No review time penalties from participating; and 3) Pilot submissions have defined criteria for completion. Pilot submissions need to be clearly defined in scope, expectations and completion endpoints.

Industry/Government Partnerships

It is important that registrants and their regulatory authorities work together in advancing the applications of EDS. An industry/government partnership is critical to the successful implementation of EDS because of the interdependence between the two entities. This partnership is necessary not only from the aspects of sharing responsibilities and resources for achieving results that benefit both, but more importantly to ensure that their respective computer hardware, software, procedures, and infrastructure are compatible.

National

EPA and ACPA have started a process to coordinate our respective approaches to EDS. The initial step was the joint workshop in March 2000, and follow-on workshops are also planned. EPA is looking at the experience of the FDA for utility in the pesticide registration process. ACPA agrees with the approach that EPA is taking in identifying those functions of its reviewers which EDS is believed to provide the greatest efficiency gains and verifying actual results through pilot submission projects. We believe, however, that the pilot submission phase of implementation is critical and calls for greater coordination between EPA and ACPA in defining the needs and objectives of various pilot projects.

We can then follow-up in facilitating registrants' participation in pilot submissions. ACPA's role in working with EPA should also concentrate on ensuring compatibility between the EDS systems of EPA and the registrants.

Regional

ACPA is encouraged by the commitment that the EPA and PMRA are showing for EDS implementation under the NAFTA Technical Working Group (TWG). A work group has been formed under the NAFTA TWG to begin addressing EDS applications from a NAFTA perspective. It is not expected that the NAFTA TWG will develop its own EDS system, but rather to ensure that the Canadian, Mexico and U.S. systems are compatible and support the registration harmonization objectives of the three countries.

International

The OECD Working Group on Pesticides (WGP) provides a more global forum for addressing EDS implementation within twenty nine member countries. The WGP is starting to take a role similar to that of the NAFTA TWG, but obviously will be faced with much greater challenges in ensuring that the implementation of EDS by its twenty nine member countries are standardized. The OECD WGP agreed at its February 2000 meeting to develop a concept paper on how it should organize the task of addressing EDS.

EDS Support of International Harmonization

There has been a significant increase in the amount and complexity of data needed to support registration of pesticides which has placed extensive burdens on both regulators and registrants. As a result, there is increasing interest among both national regulatory authorities and registrants to harmonize their registration processes. To this end, the efforts are underway in the cooperative government organizations such as the OECD Working Group on Pesticide, NAFTA Technical Working Group on Pesticides and the European Union. The stated goal of EPA for international harmonization is to develop common or compatible approaches to the review, registration and standards-setting processes. EPA states that more harmonized regulatory programs for pesticides will lead to: (1) improved food safety; (2) reduced regulatory burden on national governments; (3) strengthened scientific procedures; and (4) fewer trade problems.

ACPA strongly supports the international harmonization of pesticide registration processes for the goals stated by EPA and because of the substantial resource and time saving potentials for gaining additional country registrations. The overall intent of harmonization is that national governments share tasks in the review of a registration petition and mutually accept decisions on study reviews. However, actual approaches to harmonization must be based on a common set of data requirements, common guidelines or protocols for conducting experiments to fulfill each of the data requirements, and common quality control criteria for determining the acceptability of study reports. Scientific reviews by governments for each study report should be written in a common format that allows for mutual acceptance by all governments involved in the harmonized regulatory process. In addition, the actual implementation of harmonization must be based on the following fundamental principles:

- Assure the continued use of sound science in the regulation of crop protection products while preserving reasonable protection of intellectual property.
- Provide measurable cost reductions in data generation, e.g., minimization of multiple testing to meet the same data requirement of individual countries;
- Facilitate increased cooperation among regulators; and
- Reduce the time from submission of a registration petition to a final registration decision.

EDS has an important role in the international harmonization of pesticide registrations in facilitating the shared reviews among countries. Since multiple countries will be involved, this presents an expanded aspect to the issue of EDS systems compatibility.

There is increasing interest among government pesticide regulators and registrants in Electronic Data Submission (EDS) for its potential efficiency-gains in the pesticide registration process. Programs are being started at national and international levels to implement EDS. The fundamental principles that should guide these implementation programs are to: 1) find applications of the technology that are responsive to the needs of government reviewers; 2) ensure sufficient systems compatibility between government reviewers and registrants; and 3) select publicly available, proven and reliable technology that can easily integrate additional benefits through technology advances.

Chapter 20

Food Quality Protection Act of 1996: A New Challenge for Data Generation and Submission

W. T. Beidler and L. D. Bray

Syngenta Crop Protection, Inc., 410 Swing Road, Greensboro, NC 27419

The Food Quality Protection Act of 1996 has fundamentally changed the procedures used to assess the safety of U.S. food tolerances. The new requirements for aggregate and cumulative risk assessments alone create a huge data deficiency and hence an urgent need for real world data on exposure from all plausible pathways including dietary, water, and residential. Tremendous amounts of useful information are already available from various sources, but the data must first be organized into linked data bases in order to conduct assessments which are coherent in terms of the demographic, temporal and spatial factors. In many cases, the requisite data simply do not exist. The results of a market basket survey designed to fill an existing FQPA data gap will be presented. The market basket survey typifies the kind of studies that will be necessary to support pesticide registrations in the future.

The Food Quality Protection Act of 1996 is the most important and complex environmental legislation promulgated by Congress in the last twenty years. The FQPA, which amended both FIFRA and FFDCA, passed both houses of Congress unanimously with virtually no debate. On August 3, 1996, the day the President signed the Act, all pesticide tolerances immediately became subject to the more stringent safety standard and data requirements embodied in the FQPA language. At the core of the new safety standard is a fundamental change in the definition of "safe" as it relates to pesticide exposure. Under FIFRA, prior to FQPA, a pesticide registration was considered safe if the legal use resulted in no unreasonable adverse effect on the environment or to human health. Subsequent to FQPA, a pesticide use will be considered safe only if it can be established that there is a reasonable certainty that no harm will result from aggregate exposure to the pesticide chemical residues, including all anticipated dietary exposures and all other exposures for which there are reliable information.

The requirements of the FQPA will result in major changes in the future development of pesticide safety profiles. Examples of the most important and complex new requirements include: assessment of aggregate risk (diet, water, and residential exposure for one chemical), assessment of cumulative risk (combining aggregate exposures for all pesticides sharing a common mechanism of toxicity), assessment of acute exposure (early risk assessments focussed on chronic exposure only), testing for endocrine effects (direct or indirect hormonal changes), and increased safety margins between exposure and toxicity (additional safety factors to account for a possible enhanced susceptibility to children). These new requirements trigger comprehensive risk assessments for all uses and based on the outcome, additional studies may be necessary. In addition, the FQPA stipulates that all current tolerances are to be reassessed within ten years with the first third completed by August 1999, the second third in August 2002 and the remainder by 2006. The EPA was successful in completing the first round of reassessments, which resulted in approximately 1500 tolerance revocations. However, almost all of the revocations (1248) were for tolerances on unsupported uses (voluntary cancellations on the part of the registrant). Very few tolerance revocations were based upon the conclusions of risk assessments performed in compliance with the full mandate of FQPA.

A hallmark of the Clinton Administration has been the protection of the nation's children. The FQPA codified this theme by requiring that an additional ten-fold margin of safety shall be applied for infants and children to take into account potential pre- and post-natal toxicity and completeness of the data generated with respect to exposure and toxicity to infants and children. The statute goes on to state that notwithstanding such a requirement the

Administrator may use a different margin of safety if, on the basis of reliable data, such margin will be safe for infants and children.

The FQPA safety factor consideration creates a number of fundamental and serious issues for the pesticide industry, as well as the chemical industry as a whole. First, what constitutes a reliable database for making a determination of children's susceptibility and how should studies be conducted and interpreted for the evaluation of susceptibility. Second, since much of the needed information is not now available (particularly in the area of exposure) what safety factor, if any, should be applied in the interim while data are under development. Third, how does the added safety factor designated for an individual chemical translate in the derivation of safety factor tailored for an entire class of chemicals with a common mechanism of toxicity (cumulative risk assessment).

In order to address these questions, EPA established a 10X-Safety Factor task force in March of 1998. The task force recommended expansion of the core toxicology guideline requirements to include immunotoxicity studies in the rat as well as an *in vitro* system, developmental neurotoxicity, and acute and subchronic neurotoxicity studies. Also discussed by the task force in connection with fully evaluating the issue of infant toxicity were the following studies: pharmacokinetics in fetuses, direct dosing of offspring, enhanced developmental neurotoxicity tests (specialized testing of sensory and cognitive function), developmental immunotoxicity, developmental carcinogenicity, and potential endocrine induction effects. In addition to these newly required studies and the potential future studies related to hazard evaluation, there is also an urgent need for exposure data, particularly with respect to children. The number and complexity of the newly required studies and related information will certainly create a demand on the current methods of data generation, data handling, and reporting of data.

FQPA directs, as part of the reassessment process involved with the establishment, modification, or revocation of a tolerance, that EPA should consider all available information regarding the cumulative effects of pesticides that have been demonstrated to have a common mechanism of toxicity. Several independent scientific panels have concluded that the organophosphates pesticides (OPs) belong to a class of compounds which cause the same critical effect, act by the same molecular mechanism, act on the same molecular target and thereby share a common mechanism of toxicity (inhibition of acetylcholinesterase). Due to a combination of factors, including the common mechanism determination, the demonstrated acute toxicity, and the relative abundance of historical data, the OPs have been the focus of FQPA implementation efforts from the outset. Accordingly, most of the aggregate and cumulative risk assessment methodologies currently in place were developed and validated with OPs as the model.

Recognizing the need for industry to play a significant role in the development of methodology to accomplish the FQPA assessments, representatives from six companies agreed on a collaborative effort to conduct a multiple compound cumulative assessment using a small group of OPs as the case study. The chemicals chosen for the case study were chlorpyrifos, diazinon, azinphos-methyl, acephate, and malathion which collectively represent over 500 tolerances of the approximate 1500 tolerances established for all OPs. The six-company task force, known as the OP Case Study Group, has met on a regular basis throughout 1998 and 1999 with a focused effort toward advancing the science of the aggregate and cumulative risk assessment process. The Case Study divided into four sub teams which focused on the following areas of the risk process: use and usage, dietary, non-dietary, water, and toxicology. As stated above, the initial objective of the Case Study Team was to conduct an aggregate and cumulative risk assessment for the registrant's OPs. It became apparent early in 1998 that adequate methodology was simply not available for assessing risk per the FQPA mandate so the first two years of effort were devoted to method development. The key accomplishments of the Case Study subteams are discussed below.

The initial objective of the non-dietary subteam was to develop an exposure assessment software tool to estimate exposures from residential uses of pesticides. The final work product, an Excel-based spreadsheet called REx, can be used to calculate exposures to individuals applying pesticides in and around the home. The software can also calculate the exposure to adults and children re-entering pesticide treated lawns, gardens, and homes. REx can aggregate exposure from all routes (oral, dermal, inhalation) for multiple exposure scenarios and can be employed for both deterministic point estimates of exposure and probabilistic estimates of exposure when linked to a statistical program such as Crystal Ball.

The water sub team has completed two major projects. The first project involved an exhaustive survey of the existing inventory of historical occurrences of the five Case Study OPs in U.S. waters during the period 1990 to 1997. The data sources included state lead agencies responsible for enforcing the Safe Drinking Water Act compliance monitoring program, the U.S. Geological Survey (USGS) National Water Quality Assessment program (NAWQA), EPA's Water Quality Information System (STORET) water database, universities, and a literature search. The final report from this survey collated 114,529 individual analytical results for finished drinking water (27,300 samples) and unfinished/non-drinking water (87,299 samples). In addition to the retrospective analysis of historical water data, the subteam developed a protocol for the prospective monitoring of drinking water. During 1998 a multiresidue method was developed to detect the five OPs and corresponding oxygen analogs at levels of 0.050 ppb. Samples of finished water were taken weekly during pesticide use periods and monthly at other times for a period of 12 months. Community water systems (CWS) originating from surface water in areas with an OP use history were targeted for the study. A total of 44 CWSs participated in the study. It should be emphasized that all of the participating water supplies were very likely to draw from a water source exposed to runoff from agricultural and urban use of the specific OPs. The finalized study report details the results from a total of 1,115 samples analyzed for ten active ingredients and accompanying metabolites (22,300 individual analytical determinations).

The dietary subteam has explored several areas of the risk assessment process focusing primarily on development of a probabilistic cumulative acute dietary exposure estimate for five OPs on a limited number of crops. Since there is virtually no existing data on co-occurring residues on individual samples, industry currently relies upon compliance monitoring data from FDA and market basket data generated by USDA from its ongoing Pesticide Data Program (PDP). However, nearly all of the data on real world residues are for composited samples. In order to use the composite data in an acute dietary assessment, the residue information must first be statistically decomposited transforming the distribution of composite values into an estimate of the underlying single-serving residue distribution. The dietary subteam has developed a new statistical procedure called MaxLIP (Maximum Likelihood Imputation Procedure) which uses maximum likelihood estimation and Monte Carlo simulation to impute the frequency distribution of single-serving residue concentrations from a distribution of composite residue concentrations. The procedure was validated by a peach study conducted by Novartis Crop Protection. In the study, an orchard was treated with diazinon and 200 peaches were randomly sampled. Each peach was extracted with solvent and analyzed individually. Aliquots of each solvent extract were also combined to form a composite extract that was also analyzed. (The 200 peaches were randomly combined into 20 composites.). The 20 composite values were provided to the authors of MaxLIP and a blind imputation of the single-serving residue distribution was performed. The distribution of the imputed single-serving residues and the distribution of the actual measured single-serving residues matched closely.

Another inter-industry coalition was founded for the purpose of addressing specifically the issue of OPs in the food supply. The Organophosphate Market Basket Survey Task Force (OPMBSTF) was organized in 1998 for the purpose of assessing organophosphate pesticide residues in the nation's food supply. The task force (comprised of seven pesticide registrants) designed, funded, and conducted the project in collaboration with a third party Study Director, a Principal Investigator (responsible for the statistical design, data handling and reporting), and five contract analytical laboratories. Sample collection began in

January 1999 and continued for 12 months. The objective of this study was to determine the distribution and incidence of OP residues on single unit serving sizes (e.g., apples and peaches) and multiple unit serving sizes (e.g., grapes and green beans) commodities with a focus on foods commonly consumed by children. The commodities were sampled from retail grocery stores and were prepared prior to analysis as they would be prior to consumption (e.g., apples were washed and cored, cucumbers were washed with one inch removed from each end, oranges were washed and peeled, etc.).

A multiresidue analytical procedure was developed capable of detecting a total of twenty two active ingredients and twenty three toxicologically relevant metabolites. However, only those OPs with tolerances supported by a registrant on a specific crop were included in the analytical procedure for that crop. The limit of quantitation (LOQ) for each analyte was predetermined to be 1 ppb using gas chromatography with flame photometric detection (GC/FPD) in the phosphorous selective mode. The limit of detection (LOD) was statistically determined for each analytical run and for each analyte. All detects above the LOQ were confirmed by either mass spectrometry (GC/MS) or GC/FPD using a column with different polarity than the primary column. Residues between the LOD and the LOQ were designated as trace. Upon evaluation of confirmation/de-confirmation results conducted on samples with quantifiable residues, it became evident that a potentially large percentage of trace samples may have in fact been false positives, as was observed with the quantifiable residues. Consequently, a Trace Confirmation Analysis (TCA) program was incorporated into the study design to evaluate the relative occurrence of false positives among trace detects.

Twenty samples of each of thirteen commodities were sampled every two to three weeks (total of 25 shops) with a target of 500 individual samples per commodity over the one-year period. An A and B replicate was taken at each sampling, so a total of nearly 13,000 samples were coded and tracked through the final report phase. Samples of the thirteen commodities were purchased from approximately 500 different locations throughout the contiguous United States. The sampling phase of the study employed a stratified systematic sampling protocol, which was statistically designed to assure sampling from retail stores that would be most representative of the nation's food supply. The stratification took into account such factors as geographical location, urban/rural nature of the surrounding population, and the size of the retailer.

Each of the analytical contract laboratories was assigned the analytical responsibility for three or four specific commodities throughout the course of the project. An Excel spreadsheet "workbook" was developed specifically for use by all of the laboratories. The workbook was comprised of worksheets for each analyte and included entry fields for sample code, initial injection volume, subsequent dilution volumes, second column confirmation, GC/MS confirmation, and all other information necessary to correlate an individual single-serving sample and corresponding analytical results. The workbook was structured such that each parameter would only be entered once to minimize the potential for data entry errors. As data were generated, the individual laboratories completed a preliminary internal QA/QC of the results. Preliminary findings were reported to the Study Director for a "spot check" conducted by comparing verified chromatograms with the raw data. After verification by the Study Director, a locked version of the results was forwarded to the Principal Investigator, where data were uploaded into separate commodity files. During the upload, an automated QC routine was performed to check for formatting and data entry errors. Finally, master files for each commodity were developed which contained all of the residue information for all 500 samples. From the master files for each commodity, a summary file was compiled. The compilation of this summary file employed MACROS which perform summary statistics for various report formats and includes outputs like % trace, % non-detects, % greater than LOQ, minimum/maximum, average, and standard deviation. Throughout the entire course of the study an independent QA auditor conducted running QA/QC of the raw data. The auditor also performed quarterly in-life phase audits and assumed responsibility for the final report audit.

Upon completion of the project, the OPMBSTF will have generated a total of approximately 130,000 individual analytical results, stemming from nearly 250,000 analyses (i.e., dilutions, re-analyses, etc.). These data can be used to estimate the cumulative exposure to OP residues on the thirteen commodities chosen for the study as well as commodities similar in nature (surrogation). The chromatograms from just one single shop (20 samples of each commodity) constitute a stack of paper four linear feet in height and the total chromatogram volume for the entire study amounts to about 700 cubic feet. All of the data will be submitted to the EPA on CD's and include all workbooks. Hardcopies of representative chromatograms will be provided for half of one shop (ten samples of each commodity). (Electronic versions of the chromatograms will *not* be submitted to the EPA.) The paper raw data will be permanently archived in limestone caverns in Hutchinson Kansas.

The previous examples of the FQPA-inspired joint scientific efforts represent just a small portion of the total effort on the part of both EPA and the pesticide industry to fully interpret and implement the Food Quality Protection Act. As implementation proceeds, the requirements for supporting the registration of individual compounds have continued to expand as science policies evolve in the various risk assessment areas. Also, the need to generate simultaneous cumulative data on entire classes of chemistries, as in the examples for the OPs discussed above, will present a challenge for data capture and reporting in the future.

Chapter 21

Electronic State Submissions

Charles H. Koopmann

Stewart Agricultural Research Services, Inc., P.O. Box 509, Macon, MO 63552

Electronic submission of state pesticide registrations has been slow to gain acceptance at the state level. Primary reasons include; a) inability to receive and process data due to limitations of the current state databases, b) restrictive state regulations, c) costs, and d) apathy. Just as EPA scientists are reluctant to give up the paper trail, state officials are more secure in hard copy processing and filing procedures. The conversion to electronics has to prove its value and reliability before there will be acceptance. A few states are striving to be leaders in this effort and are proving that Internet publishing of information saves them time and improves service to their customers. As more registrants create labels, reports and supporting documents in transmittable electronic format there will be more pressure for states to accept such documents, including pressure to accept electronic payment of registration fees.

Just as the US Environmental Protection Administration (EPA) controls, approves, and monitors pesticide products for the federal government, there is a corresponding regulatory agency in the state government of each of the 50 states, District of Columbia and Puerto Rico. These state agencies, be they in the Department of Agriculture, Environmental Affairs, or elsewhere, are charged with protecting the citizens of their respective states from unlawful use of pesticides.

Put yourself in the position of the State Pesticide Administrator. He/She must conform with federal laws and regulations and state laws and regulations, while satisfying a clientele that includes every citizen and registrants who range from giant chemical conglomerates with hundreds of dynamic registrations to local/regional companies with one registration that never changes. Each of the 50 states, the District of Columbia and Puerto Rico have specific registration procedures, no two of which are alike. Every state application and renewal form is unique. State fees vary from zero (0) to $350 per product per year. The majority of the state registrations is good for one year and expires on December 31. However, in Connecticut products are registered for 5 years. Washington, Texas, and Hawaii register products for two years. Seven states renew products in June, one in August and one in November. And that's just a small example of the state differences.

Imagine yourself as the registrant, trying to satisfy the pressures of management and, of course, the marketing department by having products registered on "their" schedule, no matter when "they" get them to you. At the same time you are trying to satisfy the differences of 52 registration entities some of which can return your product approval within two days, some of which have a set review schedule with specific deadlines and still others that can take up to three months to a year to grant approval.

Do You Surrender?

Enter the electronic age, where computers will solve all time and data management problems. If you believe that, we have a bridge you might like to buy and ACS could have chosen a different theme for this book. Now that I have laid out a complicated and time consuming picture, lets look at where we stand as a pesticide industry in the matter of electronic state submissions. We know that demand for electronic data management activity is increasing dramatically, reaction time and capability have a long way to go, and websites are popping up all over the place.

One company, Kelly Registration Systems (KRS), had the foresight approximately 6 years ago to try to get ahead of this curve. KRS has a registrant version and a state version software that will allow registrants and states to communicate with each other electronically and confidentially. KRS's software database system calculates the individual state fees, completes and prints the individual state forms and facilitates electronic transfer of the required supporting documents in PDF file format. However, only one state

(Oklahoma) can receive the completed files electronically and only seven states will accept supporting documents in PDF format either via diskette or e-mail.

What Are the Obstacles to Electronic Data Submission?

The biggest obstacle is incompatibility at all levels. Remember the differences in federal and state laws and regulations, state forms, state fees, supporting data requirements, etc. Then add to that a multitude of computer systems (or lack thereof). The dynamics of each state's pesticide division and state computer department plays heavily on progress and change. Also, it wouldn't be government if politics were not involved. And, of course, with people involved there are turf and job security concerns.

The ideal system would allow the registrant to complete the state registration and renewal forms, attach the supporting documentation, make the payment and send the data right from their desk. On the state side they would be able to receive, review, accept and notify the registrant of their registration decision from the state office desk within 24 to 48 hours.

One real problem, just now being addressed by Congress, involves the acceptance and legal ramifications of electronic signatures and the transfer of cash in support of the applications and/or renewals. For years federal and state governments have been able to pay civilian and military personnel via a bank transfer to the payees bank. However it is still almost impossible for the "Registration Division" to receive an application or renewal without a paper check attached. The problem is lack of confidence in the payment process. The physical check means the money has been received and in some states allows them to return the check, with the application, if the application is rejected for being incomplete. Incomplete data and documentation usually cause these rejections and the return of the entire application. Change is hard to accept.

What Does it Take to Register an EPA Approved Product in All 52-State Entities?

It takes paper work, money and time (in that order). There are 52 different state forms with fees ranging from $0 to $350 per year as of June 30, 2000 (totaling $6,600 for one product in all states). Timing is critical regarding the date of receipt and acceptance in a few states. Some states will set the approval date as the date the money was received even though it may have taken them two months to process the forms. Still other states have a submission deadline to correspond with the planned meeting dates of their approval authority (Board, Commission). Still another (CA) can be even more time consuming than the EPA.

What Do You Need to Support Your Registration Application?

Fortunately, with the exception of California, you do not generally have to duplicate the large number of studies that you were required to submit to EPA. The EPA approval letter and the final EPA approved "master" label authenticates the EPA decision. Once you have the EPA approval letter, you must look at the 52 state entities individual registration requirements. These may include the completed application form (one to six copies), the "in-commerce" label (actual label or printer proof), the Material Safety Data Sheet (MSDS), and/or the Confidential Statement of Formula (CSF). In the case of California, you may be required to submit efficacy, toxicity and/or chemistry studies in support of your registration application.

Supplemental distributor registrations are handled differently. A sub-registration is obtained from the EPA with the permission of the basic registrant through the submission of EPA Form 8570-5, "Notice of Supplemental Distribution of a Registered Pesticide Product". In this case, the paper flow through EPA and the States is somewhat faster because the basic product has undergone prior scrutiny and the approval process. When sending these registrations to the states there is generally less paper to submit, but this is offset by some states requiring a CSF and/or an authorization to cite prior documents in support of your application. The basic registrant has usually supplied these supporting documents directly to the state. End result: More coordination of paper.

Once this is all done, you set back and wait for approval, denial, or requests for more information and/or justification.

As you can see, the process is one of attention to detail and differences. In spite of the differences there is some semblance of order being exerted and some improvement in information flow from registrant to state to registrant and consumers. The Internet and state websites are helping information flow since now some states are making the status of pesticide registrations available for public access.

California uses the Internet and e-mail for extensive communications. You can check registrations and search their website, www.cdpr.ca.gov, for your product registration status. In addition, you can receive a weekly e-mail message listing the products entering and exiting their evaluation process. You can also check product registration status online for Alaska, Hawaii, Idaho,

Nebraska, Nevada, New York, Oklahoma, Oregon, South Dakota, West Virginia and Washington. State registration regulations, forms, and instructions are also available online in about 20 states.

In the case of those states using the online connection of Kelly Registration Systems, at http://www.kellysolutions.com/, a linking of the State information with EPA information makes it possible to search for products by approved site or pest (or combination thereof) as well as by chemical ingredient. These are but a few examples of how states are trying to do a better job of communication in order to become more efficient.

Recently, the need to enhance Internet publication of registration data has received a boost from industry concerns about the rapid growth of e-sales of pesticides. Companies such as XSAg.com, Rooster.com, etc., indicate that they are, in fact, attempting to check for state registration approval prior to confirmation of the sale. Their activities have increased the inquiries to state registration offices. Phone call verifications are time consuming and interrupt the normal flow of office paperwork. This area has many legal bridges to cross and is already consuming considerable time at State offices. Change is happening daily.

If you want to explore added sites for more related information check out websites of the:

US EPA (http://www.epa.gov/),
Etoxnet (http://ace.orst.edu/info/extoxnet/),
CDMS (http://www.cdms.net),
KellySolutions (http://www.kellysolutions.com),
NPIRS (http://ceris.purdue.edu/npirs/), and
websites of the individual chemical and distributor companies or the websites of State Extension Offices. There are more data out there than can be absorbed. The challenge is to find what you need and understand it.

Conclusion

Even though the state registration system of pesticides may be complicated and slow to change, the dynamics of the industry, Internet communication, and e-commerce are causing states to evaluate their present systems. Uniformity will not happen, but similarity is always there. The electronic age will reduce the pressure for uniformity because data capture is easier, faster and less cumbersome than compromise. The challenge for the registrant is to know the basics, recognize similarities and work toward uniformity while at the same time complying with federal and state Laws and regulations, since the penalties for sale of unregistered products can be high.

Chapter 22

Electronic Pesticide Labeling: Creation, Submission, Review, and Dissemination

Thomas C. Harris

Office of Pesticide Programs, U.S. Environmental Protection Agency, Mail Stop 7505C, Ariel Rios Building, 1200 Pennsylvania Avenue, N. W., Washington, DC 20460

Labels of all pesticide products used in the US must be reviewed and approved by the Office of Pesticide Programs (OPP) of the US EPA. OPP handles about 6,500 label submissions each year. The content and format of labels are highly controlled. OPP is currently designing a system to replace paper labels with electronic text files of labels. Electronic text files would allow the EPA to 1) route labels more efficiently, 2) use text comparison software to identify all differences between proposed and currently accepted labels, and 3) facilitate dissemination of approved labels. The primary goal is to reduce the time it takes to review an application for pesticide product registration. Two exchange formats were considered: portable document format (PDF) and rich text format (RTF). In winter 2000 the EPA selected portable document format (PDF) as the standard exchange format. During 2001 OPP is piloting the use of PDF labeling focusing on evaluating the usefulness of the comparison tool in Adobe Acrobat to identify differences between label versions.

Introduction

Data and labeling are the two major components of pesticide registration. While most of the other chapters in this book discuss the electronic handling of *data* associated with pesticide registration, this chapter will focus on efforts to utilize electronic tools to streamline the review of pesticide *labeling*. Data are collected and analyzed mainly at the pesticide active ingredient level (there are about 900) while labeling is submitted and reviewed for each of the 20,000 currently registered pesticide products made from these active ingredients.

All pesticides used in the United States must be approved for use by the US Environmental Protection Agency (EPA), Office of Pesticide Programs (OPP). EPA is responsible for ensuring that a registered pesticide will not pose unreasonable adverse effects to humans or the environment. To make this determination, OPP uses both data and labeling to analyze the risk associated with each pesticidal use before deciding to accept or reject a product registration. Ultimately, the label is the legal tool used to enforce proper use of a pesticide product.

Since the audience for this chapter is mainly chemists familiar with the data aspects of pesticide registration, this chapter will begin with an overview of pesticide labeling followed by a description of the current (paper based) label registration process. Areas are then identified which can be improved through the use of electronic tools. Finally, this chapter will present the status of current OPP efforts to use the computer to speed label review through electronic label comparison.

Pesticide Labeling - the Basics

To register a pesticide product, a registrant (usually a major chemical manufacturer, a secondary formulation company, or a repackaging company) submits information to OPP including a proposed label, a confidential statement of formula (CSF), and required data (or citation of existing data with appropriate compensation to the originator). The risk associated with the product is analyzed by OPP as a function of its toxicity (as determined by study data) and the exposure of humans and the environment to the product (as determined by the label directions). If found not to pose an "unreasonable adverse effect" and if the label text and layout conform to current regulations then the label is stamped accepted by OPP.

The exact content requirements for a pesticide label are defined in the Code of Federal Regulations (CFR) Title 40, Part 156. A label includes the identity of the product (a product name, EPA registration number, list of active ingredients), a signal word (a human toxicity descriptor), first aid statements, directions for use (where to use the product, how to apply it, how much to use, safety precautions and use limitations), and precautionary (toxicological, physical, chemical, environmental) text. The wording, location, and even font size of much of this information is prescribed by regulation and policy. Even the remaining free style text must meet goals of clarity and completeness. To ensure that a product can be used effectively and safely OPP carefully reviews each label to determine if it meets these regulations, clearly states the directions for use, and does not contain any false or misleading statements.

Usually, the federal label (FIFRA Sec 3) submitted to and reviewed by OPP is the "master label" for a given product, i.e., the label describing all possible use patterns for a product. Occasionally, a "supplemental label" may be submitted to and approved by OPP. Supplemental labels are short labels used to avoid reprinting an entire label and which describe only new, additional use directions for a product. They must legally be used in conjunction with a full product label. Twenty years ago when labels submitted to EPA were produced either on a typewriter or a printing press supplemental labels were viewed as a resource saving technique for the registrant since only a page or two had to be redesigned, reprinted, resubmitted to OPP for review, and, when accepted, redistributed with product instead of the entire label. Since computers today make it simple to edit and reprint entire labels, the submission of supplemental labels for approval is now discouraged by OPP. However, once a new master label is approved by EPA a registrant may still use a supplemental label to implement the additional approved use(s) in the marketplace.

From a processing perspective, labels can be classified in three groups: new product labels, label amendments, and notification labels. Approximately 1,000 new pesticide products labels are approved each year. After their initial registration many product labels are amended (revised) to add or delete uses of a product and to keep up with current label regulations. OPP receives about 3,000 label amendments each year and these make up the bulk of OPP label reviews. Whereas the review of a new label must examine all aspects of the product labeling, the review of a label amendment can be focused on the incremental changes made since the last accepted label. Certain minor changes can be made to a label without OPP review as long as the registrant notifies OPP of the change. About 2,500 of these notification labels are received each year.

Two registration concepts are worth noting. First, the actual label attached to a product as sold may not have been reviewed by OPP. This "final printed labeling" must legally be equal to or a subset of the EPA registered label. However, it is fairly common for a registrant to subset their master label to target specific markets (e.g., agricultural, ornamental, and homeowner uses of the same product). This may result in a large number of these final printed labels and it is impossible to review every permutation. Subset final printed labeling is allowed without review as long as specific rules are followed concerning the formatting and minimum content of a subset label (failure to follow these rules may result in an enforcement action against the registrant). Second, the "current registered label" for a given product often exists as a concept more than a physical entity. For many products the current registered label is the sum of the current stamped accepted master label plus all stamped accepted supplemental labels plus all non-reviewed notification labels which have been submitted subsequent to the date of the current stamped accepted master label.

Current (Paper Based) Label Registration Process

The overall process of registering a pesticide product label can be broken into four broad steps: creation, submission, review (including decision making), and dissemination.

Creation

Label creation is initiated by the product's registrant. In the past, labels submitted to the EPA were either typed on a typewriter or printed by a commercial printer. Today, most labels submitted to the EPA are drafted using a computer. In discussions with registrants the most common software mentioned were Microsoft Word, Corel WordPerfect, and Quark. The first two are word processors while the last is a desktop publisher. This distinction becomes important when considering the comparison of electronic labels (see later) since word processors are inherently text based and the text is easily read by most software while desktop publishers are graphically based and the text within the graphics may not be readable by all software.

Submission

Once written, a proposed label (along with other paperwork) is submitted to OPP as a paper copy via US mail or courier. The package usually takes 3-5 days to reach the OPP mail room. When received, the OPP front-end staff in-process the submission by screening it for completeness and accuracy, entering tracking information regarding the submission in a database, and routing the package to the appropriate registration group for the main chemical in the product. This in-processing takes 3-5 days. Therefore, it currently takes 1-2 weeks for a proposed label to reach the registration staff who will review it.

Review

Label review is the most complex step. If a label involves a new use for the pesticide chemical then a risk analysis must be done. The registration staffer who receives the submission routes the label, confidential statements of formula, and all relevant data studies to the OPP Health Effects Division (HED) for human risk analysis and to the OPP Environmental Fate and Effects Division (EFED) for environmental and ecological risk analysis. This usually involves a significant amount of time to photocopy and collate the multiple packages required for review by these science divisions.

In addition to any necessary risk analysis, the text and layout of the proposed label must be compared to OPP standards as defined by 40 CFR 156, the OPP Label Review Manual, and numerous Pesticide Registration Notices. If the proposed label is an amendment to a currently stamped accepted label then the review is initially focused on the incremental changes. While a registrant will usually give a general description of what text has been modified, the only accurate way to find all the changes (which will become legally binding once approved) is to visually compare the current and the proposed labels word by word (often an arduous and time consuming task). Once identified, the impact of these changes can be assessed. Finally, all text, even previously accepted text, is compared against current labeling standards to bring the label fully up-to-date.

If new uses are determined to pose no unreasonable adverse effects and if the label text adheres to label regulations then the label may be stamped "accepted." A legal shortcut sometimes used to correct label errors is to stamp a label "accepted with comments" which means the label is acceptable as long as the changes outlined in the accompanying OPP letter are made. The practice of accepting with comments is being discouraged as OPP moves toward electronic labeling since it makes electronic comparison of labels difficult.

If risk is unreasonable then the label is rejected. However, it is often possible to reduce (manage) risk by changing the use directions and limitations, etc. Risk management techniques can be suggested by either EPA or the registrant but ultimately the registrant must alter the proposed label and resubmit it for review. The new submission is transmitted, received, routed, and reviewed again as described above.

Dissemination

Disseminating OPP's decision concerning a product registration is the final step of the general registration process. If a label is rejected then communication occurs only with the registrant. OPP writes and mails a letter to the registrant explaining why the product was rejected and what steps, if any, can be taken if the registrant wishes to resubmit the product for consideration. If a label is accepted then OPP writes and mails the registrant a letter stating so. If the label is stamped "accepted with comments" then the OPP letter also details precise changes which must be made to the label before product using the label can be shipped. As with incoming mail, outgoing OPP correspondence takes one to two weeks to reach its destination.

Accepted labels are also made available to parties interested in pesticide registration (state regulatory agencies, EPA regional offices, other Federal agencies, and even other registrants). In the past, the stamped accepted labels were photographed and published on a set of microfiche (or CD-ROM in recent years) which were available for a fee. Today, labels are scanned, captured as images in TIFF format, and made available for free on the internet at http://www.epa.gov/pesticides/pestlabels/index.htm. If "accepted with comments" then the EPA letter detailing required changes is included with the image. Copies of EPA accepted labels are also available for a fee by submitting a Freedom of Information Act (FOIA) request to the EPA.

Areas Which Can Be Improved Using Electronic Tools

There are numerous opportunities to improve both the speed and quality of pesticide product registration utilizing the ever evolving tools afforded by computers. While current OPP efforts are focusing on just a few of these, it is important to keep the bigger picture in mind when designing an overall electronic system.

Creation

The creation of labels is already efficiently handled on the computer using word processors and desktop publishing software. However, several steps can be taken to facilitate the creation of an entirely electronic registration submission package. OPP has already made the forms associated with pesticide registration available via the internet in addition to the paper copies which can be requested by mail. Currently, the electronic forms may be filled in on the computer but they must still be printed on paper, signed manually, and mailed to OPP. Adoption of standards for electronic signatures would open the door to full electronic processing of forms. Finally, as discussed elsewhere in this volume, the adoption of data format standards can facilitate the submission and review of studies required for pesticide registration.

Submission

Electronic transmission is the obvious tool to hasten delivery of registration packages to OPP. Delivery via email or internet would effectively eliminate the 3-5 days required for physical delivery but would have little effect on receipt tracking and screening. A key issue with electronic transmission is security since product registration submissions contain confidential business information (CBI).

Review

Electronic tools can have the biggest labor-saving impact during the review process. First, electronic data and electronic labels would eliminate the need for photocopying and collating thus making it easier to route proposals for new uses to the OPP science divisions. Second, electronic cutting and pasting from data studies and labels would make it easier to write study reviews and risk analysis documents. Third, computer software can be used to compare the proposed version of a label to the previously accepted label and thus identify exactly what text has changed. This last improvement is the focus of a current OPP workgroup and is discussed in detail in the next section.

Dissemination

Transmission via electronic mail or the internet would allow faster dissemination of regulatory decisions to the registrant as compared to physical mail. As already mentioned, images of accepted labels are now available for free on the internet; this has eliminated the need for the previous CD-ROM subscription service. A further improvement would be to enable search and retrieval of labels by content as well as registration number. This could be done by linking the labels to the EPA tracking and descriptive databases and/or by creating a full text index of the labels. The accuracy of a full text index would be affected by both the amount of graphics versus text on a label and by the accuracy of optical character recognition (OCR) software if required to convert images (e.g., the current online labels) to text prior to indexing.

Current EPA Effort: Electronic Label Comparison

The current focus of the OPP electronic labeling effort is to identify software which will flag the precise differences between the text of a proposed label and that of the last accepted label (or regulatory text standards). Electronic label comparison can increase reviewer efficiency, improve the quality of label reviews, and reduce the overall registration turnaround time from submission to decision. The same comparison software could be used by registrants prior to submission to verify that only the changes they intended were made to a label thus providing an additional level of pre-submission quality control.

Enabling electronic label comparison requires OPP to set standards for both incoming label files as well as the comparison software tool. To encourage adoption and to be successful, these standards must work with any software currently in use by registrants, be easy for EPA review staff to use, fit into an overall system for electronically processing pesticide registrations, and be based on software and hardware that will be available for the foreseeable future. The last point is important since OPP actually piloted electronic label comparison several years ago but the comparison software chosen (Docucomp) went off the market. Based on these criteria two input formats have been identified: PDF and RTF.

A "portable document format" (PDF) file is a device independent print file of a document. While files are usually converted to PDF using Adobe Acrobat the format is an open architecture standard approved by the American National Standards Institute (ANSI) and can be created by other software as well.

A "rich text format" (RTF) file is a generic word processor format that handles not only the text of a document but also some formatting (underline, bold, table structure, etc.). It was developed as an interchange format between word processors. The ability to create RTF files is already included with most word processing software.

Unfortunately, there is not a clear choice between PDF and RTF. Each has its strengths and weaknesses. PDF files can be created from any software, can be internally locked for security, and can accept electronic signatures. PDF has become the *de facto* choice for electronic document systems based on documents which have been created by a variety of software. However, the document comparison tool available through Adobe Acrobat is primitive (version 4.05c was tested); it is difficult to interpret and often results in false positives. Adobe Acrobat comparison works by comparing each document (original, revised) to each other and circling areas on each individual document where changes have occurred. The user must visually examine both documents in the vicinity of circled text to determine if the changes are additions or deletions. This is somewhat time consuming but faster than visually comparing all the text in two versions of a document. False positives (i.e., circles where no text differences occur) are fairly common. In normal comparison mode Adobe Acrobat compares not only text but also fonts, spacing, and layout. While a menu option is supposed to make it possible to limit the comparison to just the text, this feature does not work well and the software often continues to identify changes due to formatting changes (which are usually irrelevant from a pesticide labeling point of view).

RTF files, on the other hand, can be imported into Corel WordPerfect (EPA's current standard word processing software). WordPerfect includes (as do most wordprocessors) an excellent text comparison tool which takes two versions of a document, merges them into a composite document, and uses strikeout and redline editing marks to indicate where deletions and additions (respectively) have occurred. This approach results in a comparison document which is intuitive to understand and thus superior to Adobe Acrobat's comparison marks. However, RTF files have critical limitations from an electronic system point of view. By definition, RTF files can only be created by text-based wordprocessors, not graphic based desktop publishers and thus cannot be used by some current registrants. Furthermore, advanced formatting, such as tables, does not always convert neatly. While usually readable, it is common for table text to extend outside the lines of a table. RTF files cannot be internally locked; external software would be required to provide security. Likewise, external software would be required to add an electronic signature to a RTF file.

In the winter of 2000, the EPA decided to adopt PDF as the standard interchange format for incoming labels. While it remains to be seen if use of Adobe Acrobat's document comparison software will result in a reduction in registration review time, the PDF format provides clear advantages over RTF regarding file conversion and security and is, therefore, a better choice as the basis of an overall electronic document system. A pilot with volunteer registrants and EPA staff was begun in February 2001 to further test the usefulness of the Adobe Acrobat comparison tool and to fine tune any standard settings required to prevent comparison problems. As this chapter goes to press Adobe Systems is issuing a new version of Acrobat (version 5.0) which may improve the comparison capabilities of the software.

Summary

The overall process of registering a pesticide product label can be broken into four broad steps: creation, submission, review (including decision making), and dissemination. OPP is developing an electronic system to improve both the speed and quality of pesticide product registration utilizing the ever evolving tools afforded by computers. Initial efforts have focused on identifying a standard exchange format and identifying document comparison software. The exchange format enables the originators and reviewers of labeling to use different software. The comparison software enables differences between versions (e.g., current accepted and proposed) of a label to be quickly identified (the first and often time consuming step in reviewing a label amendment). The primary goal is to reduce the time it takes to review an application for pesticide product registration.

When this chapter was originally written, two possible standards had been identified (PDF and RTF) but no choice had been made. There are strengths and weaknesses to each. The obvious choice as the basis for an overall electronic system is the PDF format since it can be created from almost any input software, includes internal locking functions for file security, and accepts electronic signatures. However, Adobe Acrobat performs poorly as a document comparison tool: it is difficult to interpret and often results in false positives. The RTF format, on the other hand, allows for use of the easy to understand document comparison tool available in most word processors. However, use of RTF would require additional software to handle issues key to an overall electronic system, such as, security and electronic signatures.

In the winter of 2000, the EPA adopted PDF as the standard exchange format. Pilot testing was begun to determine the usefulness of the Adobe Acrobat comparison tool and to fine tune any standard settings required to prevent comparison problems. The release of a new version of Adobe Acrobat may affect the results.

Chapter 23

Proposed Revisions to Product Chemistry Data Requirements for Registration of Pesticide Chemicals

Sami Malak and Deborah McCall

Office of Pesticide Programs, U.S. Environmental Protection Agency, Mail Stop 7505C, Ariel Rios Building, 1200 Pennsylvania Avenue, N. W., Washington, DC 20460

This chapter focuses on product chemistry data requirements for registration of pesticide chemicals. The table of requirements on page 18 of the Harmonized OPPTS Test Guidelines, Series 830, Product Properties (1996) is discussed. The table groupes all requirements into Group "A" pertaining to product's identity, composition, and analysis; and group "B" pertaining to the physical/chemical properties. The table lists the new OPPTS Harmonized Test Guidelines, Series 830 Reference Numbers (GRNs) and the old OPP numbers of 1982. One column in the table lists the requirements for registration of a technical grade active ingredients (TGAI), whereas the last two columns list what appears to be the requirements for non-integrated manufacturing-use (MP) and end-use (EP) products. Revisions to the table of requirements and the guidelines are proposed including definitions, classification and data requirements for integrated products, and assertion of the self-certification program stipulated in PR Notice 98-1.

Introduction

The congressional mandate of The Federal Insecticide, Fungicide, and Rodenticide Act (FIFRA) and the Federal Food, Drug, and Cosmetic Act (FFDCA) as amended by the Food Quality Protection Act (FQPA) of August 3,1996 *(1)* gives the Office of Pesticide Programs (OPP) of the Environmental Protection Agency (EPA) the authority to regulate pesticide, uses, storage, disposal, and transportation in the United States. FIFRA was enacted in 1947 and was last amended by the FPQA in 1996, which added the FFDCA. Since 1947 pesticide regulations and tolerances were administered by the United States Department of Agriculture (USDA) and the Food & Drug Administration (FDA) until this authority was transferred to the recently created EPA in 1970. The EPA has modified the Code of Federal Regulations (CFR) Titles 21 and 40 *(2, 3, 4)* and published several series of guidelines with the last revision in 1996 entitled "OPPTS Test Guidelines"*(5, 6, 7, 8)*. The Guidelines are augmented by Pesticide Registration Notices (PR), Federal Register Notices (FR), The Blue Book *(9)*, The Label Review Manual/Consumer Labeling Initiative *(10)*, Standard Operating Procedures (SOPs), and Standard Evaluation Procedures (SEPs). OPPTS Test Guidelines, Series 830, Product Properties (6), was developed through a process of harmonization that blended the testing guidance of the Office of Pollution Prevention and Toxics (OPPTS) and the Organization for Economic Cooperation and Development (OECD). All EPA publications can be downloaded from the internet at: http://www.epa.gov/pesticides.

What Is Product Chemistry?

Product chemistry is the science that elucidates the chemistry, identity, composition, analysis and properties of pesticide products. Pesticides can be classified into conventional (including antimicrobials), biochemical and microbial pesticide products, which are reviewed by the following divisions: (a) *Registration Division (RD)* - Registration of conventional pesticides intended for food and non-food uses; (b) *Antimicrobials Division (AD)* - Registration of antimicrobial pesticides; (c) *Biological and Pollution Prevention Division (BPPD)* - Registration of biochemical and microbial pesticides; (d) *Special Review and Reregistration Division (SRRD)* - Reregistration of pesticides registered prior to November 1, 1984, also pesticides older than 15 years after expiration of the exclusive use period: and (e) *Health Effects Division (HED)* - Reviews submissions of technical grade active ingredients (TGAI) undergoing reregistration.

Why Is Product Chemistry Needed?

The regulatory data requirements for product chemistry are outlined in 40 CFR §158.150 to §158.190 for chemical pesticides, §158.690 for biochemical pesticides, and §158.740 for microbial pesticides. Detailed guidance on how to conduct these studies are outlined in OPPTS Test Guidelines Series 830, Product Properties (*6*); OPPTS Test Guidelines, Series 880, Biochemical Test Guidelines (*7*); OPPTS Test Guidelines, Series 885, Microbial Pesticides Test Guidelines, Overview of Microbial Pest Control Agents (*8*), and OPPTS Test Guidelines, Series 810, Product Efficacy (*5*). Listed below are some of the regulatory requirements surrounding the need for product chemistry:

1. Identification and characterization of each ingredient in pesticide products.
2. Risk assessment.
3. Environmental Fate assessment.
4. Reentry determination and labeling precautions pertaining to worker protection.
5. Labeling directions pertaining to tank mixes and spray applications.
6. Labeling ingredient statement, precautionary statements, the physical or chemical hazards statement, and storage and disposal statement.
7. Expressing product's composition and some properties on the Confidential Statements of Formula (CSF).
8. Developing the Reregistration Eligibility Documents (REDs).
9. Public inquiry: chemical spills, injuries to the flora and fauna, uses, contamination to various compartments of the environment, still birth, drift, degradation, efficacy, fish kill, leaching, runoff, marketing, storage, disposal, transportation, analytical methods, flammability, corrosivity, and explodability.

Product Chemistry Data Requirements for Registration of Pesticide Chemicals

OPPTS Test Guidelines, Series 830 (*6*) groups product chemistry data requirements into Group "A", which pertains to product identity, composition, and analysis; and Group "B", which pertains to physical/chemical properties. The table of requirements cited on page 18 of the guidelines lists OPPTS Harmonized Series-830 Guideline Reference Numbers (GRNs) and the old OPP numbers. One column in the table lists the requirements for registration of a technical grade active ingredient (TGAI), whereas the last two columns list what appears to be the requirements for non-integrated manufacturing-use (MP) products and end-use (EP) products.

With the exception of TGAIs, the table of requirements, however, does not address specific requirements for the remaining integrated products, which are: pure active ingredients (PAI), MP/EP technicals, MP/EP manufactured from registered or non-registered sources or sources of an unknown composition to the EPA, MP/EP formulated from non-registered sources or sources of an unknown composition to the EPA, and MP/EP formulated from registered sources in which a chemical reaction took place resulting in the formation of new ingredients. Further, the table cites three methods under different GRNs on partition coefficient and two on water solubility as if they are three and two requirements, respectively. Also, the table cites two requirements from the Environmental Fate Guidelines, Series 835 (UV/VIS absorption and particle size, fiber length, and diameter distribution), where particle size is not listed in the index and inadvertently omitted the requirement for solubility in organic solvents. Future upgrades to the table of requirements and the guidelines are proposed in this chapter.

Proposed Upgrades to the Table of Requirements

These proposed revisions would make it easier for the regulated pesticide industry to follow regulations and comply with the specific requirements. The proposed revisions are intended to address the many issues cited in this chapter, the answer to which cannot be found in the regulations. Examples: what are the requirements for a product manufactured or formulated from non-registered sources or sources of an unknown composition to the EPA? Similarly, what are the requirements for a pure active ingredient (PAI) and MP/EP technicals? The proposed revisions were developed while working on the self-certification program and the release of PR Notice 98-1 on "self-certification of product chemistry data." In that notice, the agency introduced a new term, "non-integrated pesticide products", defined as "manufacturing-use and end-use products formulated from registered sources with no intended chemical reaction." The agency regards this as a step in the right direction toward classifying all products into two major classes: "non-integrated" accounting for approximately two thirds of all pesticides, and "integrated" accounting for the other one-third. This classification will focus our attention on re-structuring the table of requirements and propose a definition, classification, and requirements for integrated products as shown in Table I. The proposed revisions to the table of requirements are as follows:

Table I. Proposed Classification and Product Chemistry Data Requirements for Integrated Pesticide Products

Number and Type of Each Class		Data Requirements
1	TGAI	Listed in the table of requirements
2	PAI	Same as for the TGAI using the PAI as the test substance
3 & 4	MP/EP that are TGAIs	All the requirements where applicable[1]
5 & 6	MP/EP manufactured from registered sources	All the requirements where applicable[1]
7 & 8	MP/EP manufactured from non-registered sources	All the requirements where applicable[1]. Carryover impurities must be identified on all the CSFs of the data owner
9 & 10	MP/EP manufactured from sources of an unknown composition to the EPA (that does not permit its inspection by the Agency under FIFRA sec.9(a) prior to its use in the process)	All the requirements where applicable on each MP/EP and on each source[1] in the product prior to manufacturing or by isolating each source. If the sources can not be isolated, data are required on the practical equivalent to the TGAI of each source. Carryover impurities must be identified on all CSFs of the data owner
11 & 12	MP/EP formulated from non-registered sources	Same as for non-integrated products only if the full data on each source were submitted to the Agency and found adequate. Carryover impurities must be identified on all the CSFs of the data owner

Table I. *Continued*

13 & 14	MP/EP formulated from sources of an unknown composition to the EPA	Same as for non-integrated products plus the full data requirements on each source in the product prior to formulation or by isolating each source. If the sources can not be isolated, data are required on the practical equivalent to the TGAI of each source. If known, carryover impurities must be identified on all CSFs of the data owner
15 & 16	MP/EP formulated from registered sources in which a chemical reaction took place resulting in the formation of new ingredients	All the requirements where applicable[1]

[1]GRNs 830.1550 to 830.7950 using the test substances recommended in the table of requirements, also refer to the conditional requirements listed as footnotes to the Table in 40 CFR §158.190.

1. Delete the column citing outdated OPP guideline numbers.
2. Combine three methods for partition coefficient in one requirement (GRNs 8570.7550, 830.7560 & 830.7570).
3. Combine two methods for water solubility in one requirement (GRNs 830.7840 & 830.7850).
4. Include a test for solubility in organic solvents (GRN-xxxx).
5. GRNs 830.7050 (UV/VIS absorption) and 830.7520 (Particle size, fiber length, and diameter distribution) should be moved to Environmental Fate Guidelines, Series 835. The former pertains to photodegradation and the latter pertains to drift to non-target organisms, both are Environmental Fate requirements. GRN 830.7520 was inadvertently omitted from the table of requirements.
6. Designate the last two columns for non-integrated products. The case-by case for submittal of samples (GRN 830.1900) can be explained as such: "in situations of interference and/or chemical reaction resulting in the formation of new ingredients" to include: preliminary analysis (GRNs 830.1700), and modified or new enforcement analytical methods (GRN 830.1800).
7. Delete the requirement for GRN 830.1650 (formulation process) for manufactured products.
8. Delete the requirement for GRN 830.1620 (production process) for formulated products.
9. Delete the requirement for GRN 830.1670 (discussion of formation of impurities) for non-integrated products.

Proposed Upgrades to the Guidelines

1. *Integrate the self-certification program stipulated in PR Notice 98-1*:
   This program is voluntary and is intended to simplify and accelerate the processing of applications for registration and reregistration and saves paperwork and manpower to the agency and the regulated pesticide industry while maintaining protection of public health and the environment. The PR Notice directed registrants/applicants to submit an abstract summary of the physical and chemical properties of non-integrated pesticide products on EPA Form 8570-36 (Attachment-1 to PR Notice 98-1). The full data can be retained in their files to be submitted upon request. A self-certification statement, EPA Form 8570-37, must be signed and dated by the applicant/registrant certifying that the submitted information was conducted in full compliance with the regulations (Attachment-2 to PR Notice 98-1). The PR Notice applies to applications for registration and reregistration of manufacturing-use and end-use products of chemical, biochemical and

manufacturing-use and end-use products of chemical, biochemical and microbial pesticide products produced by a "non-integrated formulation system." To be eligible for self-certification, a product must be formulated from registered sources(s) with no intended chemical reaction. The definition "non-integrated products" has promoted our re-thinking on "the nature and the specific requirements for integrated products" as proposed in this chapter (Table I). Applicants/registrants not complying with PR Notice 98-1 must submit the full data requirements of the physical/chemical properties of all products.

2. Add a definition for non-integrated products, first defined in PRN 98-1 (*11*): "formulated products from registered sources with no intended chemical reaction."
3. A definition for pure active ingredient (PAI): equals to or more than 99% pure?
4. A definition for insolubility in water: less than 1.00 ppm?
5. Allow ranges for the pH (within 2 units, e.g., 7-9) and density (not to exceed 25% of the lower value, e.g., 40 to 50 lb/ft^3).
6. A definition and guidance on repacks: "no data except when diluted."
7. Explanation of the nominal concentration (see explanations in this chapter).
8. Specificity in the following terms: integrated vs. non-integrated products, manufactured vs. formulated products; TGAI vs. PAI; and nominal concentration vs. percentage by weight.
9. Chemical formula change: a change in manufacturing and/or site data requirements; and a change in formulation and/or site data requirements.
10. Method accountability for manufactured products ≥98% (*12*).
11. Concentrations for inert ingredients to be expressed as percentage by weight (no nominals because each is composed of multiple ingredients).
12. Wider limits for the ingredients can be accepted if explained as per the regulations of 40 CFR §158.175(c).
13. For consistency with the regulations in 40 CFR §158.150, delete the requirements for lower limits for impurities.
14. *Specific guidance to the GLP requirements:* [a] six studies require full compliance with the GLP standards of 40 CFR §160.135(a) which are: preliminary analysis, stability, partition coefficient, aqueous and nonaqueous solubility, storage stability, and vapor pressure - GLP and quality assurance statements are required; [b] the remaining studies in Group "B" require partial compliance with the GLP standards of 40 CFR §160.135(b) - a GLP and quality assurance statements are required; [c] except for the enforcement analytical method, the remaining studies in Group "A" are descriptive studies, erroneously included as requiring partial compliance, require no compliance with the GLP standards; and [d] proposed

the guidelines: method validation, accuracy, and precision must be conducted by a GLP laboratory performing any or all of the following three studies: preliminary analysis, stability, storage stability. It should be noted that analytical methods are required for all integrated products and on a case-by-case situation for non-integrated products (where there is interference and/or chemical reaction resulting in the formation of new ingredients. No requirements for validated published methods.
15. Aged and number of samples needed for analysis of halogenated dibenzodioxins, dibenzofurans; hexa & penta chlorobenzene; anilines; hydrazines; sodium nitrite; list 1 inerts; and others of toxicological concern similar to the recommendations for nitrosamines on page 9 of the guidelines.
16. Low level of quantitation (LOQ), similar to those in 40 CFR 766.27 (*3*), level of acceptability for ingredients of toxicological concern and what should be listed on pesticide labels?
17. Upgrade GRN 830.1700 (preliminary analysis) to explain (a) What is required for non-integrated products?; (b) What is meant by "a statement of the composition of the practical equivalent to the TGAI?" Does it mean the acid equivalent? How about production processes that do not involve an acid/base neutralization?; and (c) if the statement of composition is satisfied, should we ignore the remaining requirements? Furthermore, the statement of composition submitted by a formulator using a source of an unknown composition to the EPA is not adequate as that submitted by the manufacturer. It is the manufacturer who knows the exact composition of his/her product, particularly where components of toxicological concern may be present.
18. Re-define integrated products: "all products except non-integrated."

Highlighting the Differences Between TGAIs, PAIs, MPs, and EPs

Technical Grade of Active Ingredient

- A material containing an active ingredient that will prevent, destroy, repel, or mitigate any pest or act as a plant growth regulator, defoliant, desiccant or nitrogen stabilizer. TGAI can be MP or EP;
- Produced on a commercial or pilot-plant production scale; and
- Contains no inerts, except those used for purification. Products involving chemical reaction can be represented by the following equation: reactants A + B $\rightarrow$ TGAI, composed of PAI (nominal concentration) + residuals from the starting materials + impurities + contaminants + side reactions +

degradation products + residuals of solvents used for purification (must be claimed on the CSF as residual solvents).

Pure Active Ingredient

- It is the purest form of a pesticide with a purity equals to or more than 99%.

Manufacturing-Use Product

- Any pesticide product other than an end-use product;
- A labeling stating "for use in formulating other products"; and
- May contain solvents and/or stabilizers.

End-Use Product

- Pesticide product whose labeling includes directions for use; and
- Labeling does not state that it be used in formulating or manufacturing other products.

How to Express Nominal Concentrations and Upper/Lower Certified Limits on the Labels and CSFs

According to the "Standard Certified Limits" of 40CFR§158.175(b)(2), where N = percentage nominal concentration for manufactured products, and percentage by weight for formulated products:
$N \leq 1.0\%$..... $\pm 10.0\%$; $1.0\% < N \leq 20.0\%$..... $\pm 5.00\%$; $20\% < N \leq 100.0\%$.... $\pm 3.0\%$

The Ingredient Statement on the Label of a Formulated End-Use Product

Label (refer to Table II):
Active ingredient..24%
Other ingredients..76%
Total..100%

Table II. *Columns 13 and 14 of the CSF; Source Purity = 96%*

Column 13		Column 14	
(a) Amount in lbs	(b) % w/w	(a) % Upper	(b) % Lower Limit
250 TGAI (240) PAI	25 (24)	(24.72)	(23.28)
350 Inert	35	36.05	33.95
400 Inert	40	41.2	38.8
1000	100		

Explanation about Table II and the Nominal Concentration/Percentage by Weight Concept

- If the label claim nominal concentration is 24% and chemical purity of the TGAI is 96%, then by calculation, a formulator will use 250 pounds in a 1000 pounds batch = 25% w/w [(250 ÷ 1000) X 100]. Material balance of 100% is achieved by adding two inerts totaling 750 pounds. To calculate the nominal concentration, multiply percentage by weight by chemical purity then divide by 100 [(25 X 96) ÷ 100] = 24%. The general formula is: $N = [(P \times \% \text{ w/w}) \div 100]$, where N = nominal concentration, P = chemical purity of the TGAI, and w/w = percentage by weight. On the other hand, the percentage by weight can be calculated by dividing label claim nominal concentration by chemical purity then multiplying by 100 [(24 ÷ 96) X 100] = 25%; and
- Inert ingredients are composed of a mixture of various components, sometimes more than 100, each with a unique chemical name and CAS registry number. Therefore, it is not practical to list nominal concentrations for inerts since nominals are based on pure substances. Only if an inert is 100% pure, then a % w/w would equal to % N. However, it will be confusing to list % w/w for some and % N for others. Therefore, the percentage by weight of all inerts are listed in Column 13(b) of the CSF and their upper/lower limits can be calculated based on percentage w/w, to be listed in Columns 14(a) and 14(b) of the CSF, respectively.

The Ingredient Statement on the Label of a Technical Grade of Active Ingredient (Manufactured Product)

Label (refer to Table III):
    Active ingredient.. 96%
    Other Ingredients.. 4%
    Total... 100%

Table III. *Columns 13 and 14 of the CSF; Source Purity = 96%*

Column 13		Column 14	
(a) Amount, lbs	(b) % (nominal)	(a) % Upper Limit	(b) % Lower Limit
960 PAI	96.0	99.0	93.0
15 impurity	1.5	2.0	1.4
20 impurity	2.0	2.5	1.7
995	99.5		

Explanation about Table III and the Nominal Concentration's Concept for a TGAI

Based on sample analysis using the enforcement analytical method, as shown in Table III, the purity of the TGAI is 96% (nominal concentration) and the nominal concentration of two impurities are 1.5 and 2.0% for a total method recovery of 99.5% equivalent to 995 pounds. Because batch production or the actual yield was 1000 pounds, that means 5 pounds or 0.5% of the amount was not accounted for by the method for various reasons associated with the analytical method, purity of the reactants, laboratory conditions, equipment, reagents, analytical chemists, and technicians. Method accountability ≥98% is acceptable (*12*). Please note that the lower limits are required by the guidelines but not by 40 CFR §158.

Explanations of the Nominal Concentration

The nominal concentration is defined in 40CFR§158.153, copied in PR

Notice 91-2 and OPPTS Test Guidelines, Series 830-Product Properties as "The amount of an ingredient which is expected to be present in a typical sample of a pesticide product at the time the product is produced, expressed as a percentage by weight." The term "at the time the product is produced" is not a reflection of marketing realities, because it may take several years from production before consumers use the product. Was the product efficacious when used? Or, was there a need for a labeling expiration date as per the regulation of 40 CFR §156.10(g)(6)? It is, therefore, appropriate to replace the term "at the time the product is produced" with "from production to use." Further, the percentage by weight of the TGAI is calculated from the amount based on the total formulated product, whereas, the equivalent amount to the PAI is calculated from the percentage nominal as determined by validated analytical techniques, which is the same as the label claim nominal concentration. It is apparent that it is the percentage, not weight or amount, because a manufacturer does not weigh a PAI and impurities, rather, they are components of a manufactured product. Their weights and amounts are calculated based on percentages that may or may not add up to the exact production volume (yield) and/or method recovery. Lower yield of manufactured products is not uncommon, reported in one product at 17%. As such, the definition can be modified to substitute "expressed as a percentage by weight" with "and its equivalent by weight." On the other hand, there is no need for "and its equivalent by weight" because of the expected variations in the calculated amounts due to variability in the actual or calculated yield and method accountability within one batch and among succeeding batches, given that method recovery $\geq 98\%$ is permitted by the regulations (12). It is obvious that the amount column is irrelevant to reviewing the CSF since the exact production volume is reported to the EPA for the purpose of a benefit/risk assessment. Whether the amount column is retained or deleted, the term "and its equivalent by weight" is not necessary. Finally, it is anticipated that an enforcement analytical method may not determine the exact label claim nominal concentration, rather, a value between the lower and upper certified limits, referred to as "guarantee." Therefore, a more precise interpretation for the "nominal concentration" would be:"The percent of a pure ingredient within the minimum and maximum guarantee that is most likely to be present in a typical sample of a pesticide product as determined by validated analytical techniques from production to use."

Helpful Tips to Registrants/Applicants

1. Follow the regulations and consult with the product managers who may arrange for a meeting with Agency scientists if there are questions.

Electronic contacts can be made to the Registration Division at: http://www.epa.gov/pesticides.
2. Submit complete data packages containing the necessary studies in support of your request. Do not submit superfluous information.
3. Ensure: (a) compliance with PR Notice 86-5 in formatting of data submissions; (b) that the submission is accompanied by a cover letter and the necessary forms stating specifically what is requested; (c) that the appropriate boxes on the application Form (EPA Form 8570-1) are complete, also on the same Form for "me-too requests", cite the name and registration number of the product claimed to be similar to the applicant's product; and (d) consistency between the label and CSF in citing the ingredient statement and the nominal concentrations.
4. On the CSF (Rev. 8/94): (a) indicate the nominal concentration between parenthesis below the percentage by weight in column 13b and the corresponding upper/lower limits in columns 14a and 14b; (b) indicate the purity of the source product in column 10; (c) list the same label claim nominal concentration for a TGAI in column 13b; (d) list, where applicable, the flash point/flame extension/any flash back in box 9, noting that the flash point should be determined on the base product before packaging and adding a propellent, or by degassing if a propellent was added; and (e) achieve a material balance of 100% in formulated products and equals to or more than 98% accountability in manufactured products (*12).*

References

1. The Federal Insecticide, Fungicide, and Rodenticide ACT (FIFRA) and the Federal Food, Drug, and Cosmetic Act (FFDCA) As Amended by the Food Quality Protection Act (FQPA) of August 3,1996.
2. Code of Federal Regulations, Title 21.
3. Code of Federal Regulations, Title 40, Parts 158.150 to 158-190.
4. Code of Federal Regulations, Title 40, Parts 766.27.
5. OPPTS Test Guidelines, Series 810, Product Efficacy.
6. OPPTS Test Guidelines, Series 830, Product Properties, EPA 712-C-96, August, 1996.
7. OPPTS Test Guidelines, Series 880, Biochemical Test Guidelines.
8. OPPTS Test Guidelines, Series 885, Microbial Pesticides Test Guidelines, Overview of Microbial Pest Control Agents.
9. The Blue Book: General Information On Applying For Registration of Pesticides in The United States, 1992.
10. Label Review Manual/Consumer Labeling Initiative, 2000.
11. Pesticide Regulations 98-1, entitled "Self-Certification of Product Chemistry Data."
12. Federal Register Notice: 49(207)FR42863,24/OCT/1984

Chapter 24

Electronic Data Submission: Pilot Efforts in the Office of Pesticide Programs, U.S. Environmental Protection Agency

Kathryn S. Bouvé

Office of Pesticide Programs, U.S. Environmental Protection Agency, Mail Stop 7502C, 1200 Pennsylvania Avenue, N. W., Washington, DC 20460

The Office of Pesticide Programs (OPP) has embarked on a series of pilots with pesticide registrants to develop a standard and process for accepting and reviewing electronic data submissions. OPP is seeking an electronic submission standard that strikes a good balance between the needs of registrants and data reviewers. The standard must be inexpensive and easy for the wide range of U.S. registrants to implement. It must also provide OPP reviewers with easily learned functionality that makes their work more efficient and effective. Other interests to be served include data integrity, protection of confidential business information, and international harmonization. While Adobe Acrobat 4.0 has been selected as the tool for pilot efforts, keeping pace with emerging technologies will present additional challenges. The results of early pilots as well as discussion of the many issues and concerns related to electronic submission and review will be presented.

Background

The Office of Pesticide Programs (OPP) of the U.S. Environmental Protection Agency is responsible for regulating pesticides. Its mission is to protect public health and the environment from the risks posed by pesticides and to promote safer means of pest control. OPP is essentially a licensing operation that registers pesticides for sale and use. There are approximately 20,000 U.S. registered products that include insecticides, herbicides, fungicides, rodenticides, disinfectants, sanitizers, and repellents. There are approximately 2,000 registrants and the number of product registrations they hold ranges from 1 to 541. The median is 2. Six percent of registrants hold 70% of active products.

Extensive information on the risks and benefits of using a pesticide must be submitted by applicants before it can be registered. This information includes reports of research in the scientific disciplines appropriate to the intended use of the pesticide. Applications for new uses for registered pesticides must also be presented with appropriate data. A pesticide can be registered only if it is determined that it will perform its intended function without unreasonable adverse effects on human health and the environment. That determination is made by scientists in the Office of Pesticide Programs upon review of the studies and assessment of information about hazards and exposure. Further, as new risk issues emerge, the Agency can require that additional studies be conducted and submitted throughout the life of a pesticide.

OPP determines the data requirements for various types and uses of pesticides and designs the protocols for testing the pesticides. Registrants bear the cost of conducting the studies according to the standard protocols. Full reports of the studies must be submitted to OPP along with the application for registration. In the case of pesticides intended for food uses, a petition to set a tolerance, or maximum residue level in food or feed, must also be submitted with required data.

The data requirements for pesticides vary widely depending on the proposed use of the product. At one extreme, a recent application for registration of a new food-use chemical used in an outdoor setting was accompanied by 230 different studies. Basic chemistry, acute and chronic toxicity, worker and dietary exposure, as well as environmental fate and ecological effects must be analyzed by OPP. At the other extreme, an application for a new product containing an already-registered - and well studied - active ingredient requires only chemistry data and a standard battery of six acute toxicity studies.

OPP scientists review and assess each study and prepare written documentation of their findings. During this process, basic questions are answered: Was the study conducted according to the protocol? Does the reviewer agree with the results and their interpretation presented by the study author? Does the study satisfy the data requirement it addresses? The reviewer will re-analyze the statistics and re-calculate values in tables. OPP's review document must include details about how the study was conducted and key data tables presented by the study author as well as the reviewer's conclusions. Review documents serve as the building blocks for higher tier science decision documents including hazard assessments, exposure assessments, and risk assessments. The reviews must be archived by the program to document the bases for registration decisions and they may be re-used or re-analyzed in the future.

Performance and submission of research reports by registrants and the review and assessment of those data by OPP staff comprise the essential interaction between the regulated and the regulators in the pesticide program. Finding more efficient and effective ways of conducting that 'conversation' is an important goal of the program.

Significance of Study Data in Pesticide Regulatory Program

Study data represent a major level of activity in OPP. Over the years, the program has developed guidelines for nearly 200 different types of studies for conventional, biochemical, and microbial pesticides. OPP receives approximately 8,500 studies per year. The work of over 5,500 laboratories is represented in OPP's inventory of 280,000 studies.

Because of the high volume of studies reviewed by OPP and the essential nature of the work, a significant proportion of OPP's resources are dedicated to it. Approximately 40% of staff in OPP - 325 scientists - are involved in the primary or secondary review of studies. Primary study review work is supplemented by contractor resources as well.

Registrants also face challenges as they develop and submit data to OPP. They put significant expense and effort into conducting research, preparing reports, and assembling them into application submissions that meet OPP's data requirements and formatting standards. A registrant may spend $10 to $15 Million to conduct 100 or more OPP-required studies for a food-use pesticide. Registrants also need to preserve and archive the valuable data they develop. A well-conducted study can satisfy a data requirement for many years and other registrants may cite those data in their applications provided they compensate the owner of the data. Studies may be performed over many years in many labs in different countries. Resulting reports may be developed using different software packages for word processing, spreadsheet presentation, and statistical analysis. In some cases the only available medium of a still-valid study will be hard copy.

During Fall 1999, significant factors converged that gave impetus to OPP's pilot efforts. OPP management needed efficiency improvements to address the high volume and high value of studies and their review as well as to meet growing demand on the program for regulatory decisions under statutory deadlines. OPP's technical infrastructure and staff capabilities had matured to the point where technology could be put to more sophisticated use. The program was aware of viable technology in use in industry, other EPA offices, and other government agencies. Finally, in October, 1998, Congress passed the Government Paperwork

Elimination Act (GPEA). This law mandates that by the year 2003, all federal government agencies must be prepared to accept electronic submission of whatever information they require from the public, the states, or from industry. This must be achieved using open standards and non-proprietary software and hardware to the extent possible. As a result of these factors, OPP initiated a series of pilots to test the use of Adobe Acrobat Portable Document Format and related tools as the standard for electronic submission and review of study data. It is expected to be a two-year effort because it will take time to involve a good cross section of registrants, studies, and reviewers in the pilots.

Adobe Acrobat and PDF

Before proceeding with a description of the pilots, it is important that the terminology about technology used in this paper is well understood, especially PDF, Adobe Acrobat 4.0, and Acrobat Reader. PDF is an acronym for Portable Document Format. PDF is a file format created by Adobe which enables a user to view and print a file exactly as the author designed it without needing the application or fonts used to create the file. Fonts, colors, images, and layouts are preserved. PDF files are compressed thus reducing file size, transfer time, and storage space needs. It was introduced in 1993 and Adobe describes it as an open, de facto standard.

A number of software packages are designed to save documents as PDF files, and it is likely that more will emerge in the future. Examples are the Corel WordPerfect version 9 and Adobe products such as PageMaker. At the present time, however, to create a PDF file with the enhanced features desired by OPP, the user needs the software package Adobe Acrobat. It converts word processing and spreadsheet files as well as scanned images into PDF and preserves the look and feel of the original. The user can create bookmarks and links as navigation aids and can import electronic versions of tables and spreadsheets that retain their native format. OPP specifies Adobe Acrobat 4.0 for its pilots. The software must be purchased and the price is about $240 for a single user license.

Adobe Acrobat Reader is software that permits users to view, navigate, search, and print Adobe PDF files on major computer platforms. The software is free and available from Adobe's web site. While the Reader is useful, it does not support all the functionality OPP's reviewers require. For the pilots, OPP is purchasing Adobe Acrobat version 4.0 to support review of studies submitted as PDF files. Adobe Acrobat permits review, mark up, annotation, extraction of text and tables for editing or other manipulation in addition to basic viewing, navigation, and printing capabilities.

Design and Implementation of Pilots

Adobe Acrobat and PDF emerged as the technology tools for OPP's pilots for a number of reasons. It is inexpensive. It is widely used in areas well beyond the pesticide program and can be considered a de facto standard. Adobe supports the product, seeks user feedback, and continues to increase its functionality. PDF has been selected as a standard for electronic submission of new drug applications by the Food and Drug Administration (FDA). FDA's functions are similar to OPP's and there is some overlap between the regulated communities of both agencies. OPP's sister office, the Office of Pollution Prevention and Toxics (OPPT), has selected PDF for electronic submission of data. OPP was also persuaded by the fact that PDF was suggested by several major registrants that submit large data packages. PDF lends itself to structured formats through use of a feature called "bookmarks." OPP can build on this in the future if more highly structured formats are desired such as XML. Finally, Adobe PDF is consistent with the basic requirements under GPEA.

Please note that OPP also identified the need for electronic submission of other types of program information. One is files and data that supplement studies. Another is the text of current and proposed product labels. These are topics of other chapters presented in this book.

OPP established operating principles for the pilots related to electronic submission and review of studies. First and foremost, the electronic formatting standard must strike a good balance between registrants' and reviewers' needs. It must be inexpensive and easy for registrants to adopt. It must provide reviewers easily learned functions that make their work more efficient and effective. Another principle is to build on the lessons learned by others. OPP has benefitted from the experiences of the Food and Drug Administration, the Office of Pollution Prevention and Toxics, registrants, Canada's Pest Management Regulatory Agency pilot efforts, and early electronic submission efforts in the European Union.

As OPP embarked on the first pilots, it specified what it hoped to learn from them. From the registrants' perspective, can PDF be readily produced from the variety of sources available (electronic, scanned images, etc.)? Are OPP's technical specifications clear and achievable? Can OPP ensure the integrity of the electronic version of the submission?

From the reviewers' perspective, OPP hoped to learn if Adobe Acrobat and PDF would meet a variety of requirements. Is Acrobat easy to learn? Are OPP's work stations ergonomically suitable for this type of work? How useful are the tools to review the data and to prepare review documents? Specifically, do the tools allow reviewers to navigate among studies and within a study, perform full text searches, annotate, view and print, manipulate data including exporting it to other software for further analysis, and excerpt and edit text or tables? The bottom-line question for the program is this: Is the electronically-assisted review process more efficient and more effective?

OPP established criteria for selecting candidates for early pilots. The studies should support an application for a new use for a registered pesticide. The data review must be scheduled on the current priority work plan. A small number of studies in one discipline that had already been submitted on paper seemed most manageable. The program would avoid the complication of piloting with studies that contained confidential business information (CBI). And most importantly, OPP needed a registrant ready, willing, and able to create PDF documents from the original electronic source.

In order to initiate even a simple pilot, OPP had to prepare a number of documents that would guide registrants and OPP staff on preparation and in-processing of electronic versions of studies. With the assistance of staff from the registrant Rhone-Poulenc (now part of Aventis), OPP developed technical specifications on how to prepare the PDF files using Adobe Acrobat 4.0. This document was based on specifications developed by FDA. OPP also prepared a document that specifies the process by which registrants would organize and name the PDF version of studies on compact disk and how OPP would in-process them. The in-processing includes virus scans, verifying compliance with formatting requirements, and posting the studies on the OPP LAN. It was also essential for registrants to certify that the information in the electronic version of the study was the same as the paper version of the study, so a Certification with Respect to Data Integrity was developed. Finally, OPP developed a Reviewer Assessment form that would capture the reviewer's experiences during the pilot. Topics addressed include the learning curve, the performance of the tools, ergonomics, LAN and PC performance, and the efficiency and effectiveness of the tools compared to the 'paper' process.

To undertake the first pilot, OPP staff contacted registrants that had applications and studies in house and on the priority list for review and asked if they were willing to participate in the pilot. From among them Rohm and Haas volunteered and submitted an electronic version of 7 residue chemistry studies submitted in support of a tolerance petition. OPP's Health Effects Division (HED) selected a reviewer for the data, a staff person comfortable with technology and experienced with this type of data.

The results of the first pilot were very encouraging. The reviewer found the software easy to learn in all categories of functionality that were used. The annotation feature was not used. Tools supporting navigation, text searches, viewing and printing were scored 'excellent.' Tools to support excerpting and editing text were scored 'good'. Tools to support excerpting and editing tables and the manipulation of data for export to Excel were scored 'average'. It was noted by the reviewer that a good knowledge of Excel was needed to get the data into a usable form. For this reviewer, the size of the monitor at home (17-inch) and office (21-inch) was adequate. The desk, chair and mouse posed no problems. The OPP LAN system performance and response times were acceptable.

The critical requirements of improved effectiveness and efficiency were supported by the PDF and Acrobat tools. The reviewer could better analyze the data because it was easier and faster to check data against conclusions and statistics. This was accomplished by switching back and forth between linked tables of raw data and the summaries. Regarding efficiency, the reviewer found improvement in nearly all aspects of the work. Data requiring re-analysis could be dropped into the analysis software via drag and drop rather than by retyping and error checking the data. Selected text from the study could be excerpted and dropped into the review document, with proper attribution, without loss of accuracy that may occur during the usual paraphrasing and summarizing of the information.

Suggestions for future pilots were also identified in the reviewer's assessment. For example, it would be more helpful if PDF tables were derived from true tables created by the 'table' function of the original word processing files. That suggestion was added to the technical specifications document.

Buoyed by the encouraging results of the first effort, more pilots are underway or planned. This includes entire submissions presented in PDF in early 2001. The goal is to involve more registrants, different types of data, and more reviewers - both OPP staff and contractors. It is also assumed that PDF submissions will be a mix of studies derived from electronic sources, scanned images of paper with text captured via optical character reader (OCR), and scanned images (photographs, chromatograms, etc.). OPP is eager to gain experience in all these areas.

Issues and Concerns

While pilot efforts continue, OPP must address a number of other issues and concerns in order to make electronic submission and review routine.

Communication and Coordination - Internal and External

There are other opportunities to use electronic submission tools and processes to make pesticide registration work more efficient and effective. OPP has formed the Electronic Data Submission Workgroup with representatives from all Divisions to explore those opportunities. The goal of the workgroup is to develop an electronic data submission approach that achieves operating efficiencies through the promotion and facilitation of the electronic submission process including the delivery, review, data interchange capability and archiving of data supporting national pesticide registration. This approach will be implemented using current technology, consider the needs of reviewers and stakeholders, and address legal, archival and other requirements. Sub-groups were formed which are developing and refining guidance on the overall formatting of electronic submissions and typical studies as well as supplemental files. A web site has been established to make available in one place information about OPP's pilots efforts, other workgroup activities, and guidance documents and technical specifications. The URL for the site is www.epa.gov/oppfead1/esr_pilots.htm.

A close working relationship with pesticide registrants is essential in all these efforts. Much of the progress to date on pilots and related guidance documents has been the direct result of consultation, brain-storming, and information-sharing with individuals in several registrant companies. In March 2000, the American Crop Protection Association, a trade association of major pesticide registrants, and EPA conducted a joint workshop on Electronic Submissions for Pesticide Registration. The agenda included presentations on electronic submission from industry and government perspectives, a demonstration of a PDF submission, and breakout sessions on creating and submitting overall submissions and supplemental files. The break out sessions were excellent opportunities for representatives of industry, laboratories, and regulatory agencies to critique early versions of guidance documents and specifications. The ideas and suggestions that emerged were used to revise - and greatly improve - interim guidance documents for the pilots.

Another close and fruitful working relationship is that between OPP and Canada's Pest Management Regulatory Agency (PMRA). PMRA is also hard at work on developing a standard for electronic submission and OPP has learned a great deal from their early pilot efforts. As NAFTA partners, OPP and PMRA are working to harmonize the electronic data submission and review standard and processes to the greatest extent possible. In fact, this work is being conducted under the auspices of the NAFTA Technical Working Group on Pesticides. Both OPP and PMRA are active in international harmonization efforts through the European Commission and the Organization for Economic Cooperation and Development (OECD).

OPP continues communication and coordination on electronic submission standards and processes with its sister office, the Office of Pollution Prevention and Toxics (OPPT), and EPA's Office of Environmental Information (OEI). In addition, FDA is an especially useful source of information. FDA staff have worked hard on these issues for the last several years and they generously share their knowledge and experience with OPP.

Technological Challenges

Without a doubt, the fast-changing technical environment presents great challenges to OPP and industry as they work to develop and implement an electronic data submission and review standard. While speed and change define technology today, OPP, by contrast, is a large and somewhat conservative organization. The many individuals involved in data review vary considerably by age, experience with technology, and work habits. There are approximately 25 organizational units with data reviewers in OPP. Managers of those units must implement electronic review tools consistently in terms of training and adherence to processes and procedures. The same is true for staff overseeing contract work.

OPP's information technology infrastructure must be continuously upgraded to ensure adequate work stations, sufficient LAN storage for electronic versions of studies, effective virus detection capabilities, and data security and integrity. The infrastructure must serve OPP's needs and, at the same time, remain within Agency and Federal government standards. Costs must be within the program's budget.

The registrant community faces significant challenges related to technology on several dimensions. Companies that develop and submit data are generally large, complex organizations that may face work force issues similar to OPP's. Data submitted in support of a single registration may have been developed over many years in different labs in different countries using different data collection and analysis software and word processing packages. In addition, the pesticide industry has been quite volatile in recent years with companies merging, splitting, buying and selling large blocks of products and related data. It is likely that these factors complicate efforts to establish and maintain standards for electronic archiving and submission of study reports.

All this variability within the regulator and regulated communities argues for inexpensive, simple, flexible technology solutions. At present, Adobe Acrobat 4.0 best meets those needs.

While Adobe Acrobat and PDF are a logical choice for today, this technology may be completely obsolete in five years. No where is this dilemma better illustrated than in the issues faced by the National Archives and Records Administration (NARA). Studies submitted to the pesticide program are considered permanent records according to the record disposition schedule approved by NARA. A copy of every study received by OPP is sent to NARA's Federal Record Center where it is retained for 20 years. After that, the document is transferred to the National Archives for permanent storage. At the present time the only media NARA accepts for storage of permanent records are paper and microfiche. Digitized media are not allowed. Many government agencies are acquiring and generating electronic versions of records scheduled for permanent or long term retention. NARA is tasked with finding solutions to this problem in ways that are technologically feasible and cost effective.

Conclusion

OPP will continue efforts to establish a standard for electronic submission and review of pesticide program data through a series of pilots. OPP is encouraged by the success of the early pilot efforts and by the interest and cooperation of industry and government agencies to date. The pilots will allow industry and OPP to determine if Adobe Acrobat and PDF files strike the needed balance between ease and low cost for the former and improved efficiency and effectiveness for the latter.

Chapter 25

Electronic Data Submission in the Environmental Protection Agency Office of Pesticide Programs: Supplemental Files

Susan V. Hummel

Office of Pesticide Programs, U.S. Environmental Protection Agency, Mail Stop 7509C, 1200 Pennsylvania Avenue, N. W., Washington, DC 20460

The EPA Office of Pesticide Programs (OPP) recognizes that there are significant potential benefits to be gained by the implementation of electronic submission of studies to support pesticide registration or re-registration. Supplemental files of additional study data can be submitted electronically to the EPA to facilitate study reviews and data analysis. This paper reports on OPP experiences with electronic submissions of supplemental files included as part of a full electronic submission package for a chemical, or provided separately in response to a specific need. Efforts are being made to define the nature and content of supplemental files, to define the attributes of file structure and content, and to develop guidance for their use. Issues and concerns regarding the use of supplemental files are identified and addressed and future developments outlined.

Background

In the United States, pesticide products are registered for use under the provisions of the Federal Insecticide Fungicide and Rodenticide Act (FIFRA), 7 U.S.C. 136 et seq. If the pesticide is used on foods, tolerances, or legal limits for the pesticide chemical residue in or on a food, may be required to be established under the Federal Food, Drug and Cosmetic Act (FFDCA), 21 U.S.C. 301 et seq. The Food Quality Protection Act of 1996 (FQPA) (Pub. L. 104-170) was signed into law August 3, 1996, and amended both the FIFRA and the FFDCA. The Environmental Protection Agency (EPA) Office of Pesticide Programs (OPP) is charged with administering these laws.

Under the provisions of FIFRA and FFDCA, the pesticide manufacturers (i.e., registrants), are required to submit a full and comprehensive battery of product chemistry, toxicity, residue chemistry, occupational and residential exposure, and environmental fate and effects data. All the submitted data are reviewed by Agency scientists for conformity with standard practices within the discipline and conformity with Agency Testing Guidelines. The data requirements for registration of a pesticide have been published in the Code of Federal Regulations, Title 40, Part 158 (40 CFR §158). The Agency Testing Guidelines have been published on the World Wide Web at http://www.epa.gov/OPPTS_Harmonized/. A summary listing is given below in Table I:

Table I. OPPTS Harmonized Test Guidelines

Series	Description
810	Product Performance
830	Product Properties (Product Chemistry)
835	Fate, Transport and Transformation
840	Spray Drift
850	Ecological Effects
860	Residue Chemistry
870	Health Effects (Toxicity)
875	Occupational and Residential Exposure
880	Biochemicals
885	Microbial Pesticide

The data required for pesticide registration and tolerance setting are submitted by the pesticide registrant (often the pesticide manufacturer) and the submitted data are reviewed by Agency scientists. The reviewers prepare a summary of the information and data contained in the submission, verify the data submitter's calculations, and perform statistical analyses of the data. They identify information missing in the study report and determine study conclusions. They then assess the appropriateness and adequacy of the submitted data, determine if additional information is needed from other Agency scientists, and conclude whether the submitted data support the requested action (i.e., registration, tolerance).

The technology available for both conducting the scientific studies and preparing the study report at the laboratory level, and for analyzing the data and preparing the scientific reviews at EPA has changed over the years, as shown in Table II. Initially in OPP, science reviews were written by hand, and then typed by a secretary. Calculations using the data were performed using a desk calculator or by hand. Today, almost all EPA science reviewers prepare their own reviews using a word processor, and many use other software tools for analyzing the data, such as spreadsheets, database programs, and statistical software packages. Additionally some types of data such as tumor incidence data undergo rigorous specialized statistical analyses by Agency experts.

Table II. Technology in Use at EPA

Technology Available	*Time Frame*	*Review Preparation*
Typewriter	pre-1975	Handwritten by reviewer, typed by secretary
Typewriter with memory	1975	Handwritten by reviewer, typed by secretary
Dedicated word processor	1979	Handwritten by reviewer, typed by secretary
	1981	Some reviewers typed
Computer	1984	Some reviewers typed
	1993	Almost all reviewers typed

Definition of Supplemental Files

A workshop to discuss electronic data submission in support of pesticide registration was conducted by the American Crop Protection Association (ACPA) and EPA in March 2000. The participants were divided into four groups - two groups discussed the overall data submissions, the remaining two groups discussed supplemental files. In the groups discussing supplemental files, representatives from EPA and from the pesticide industry defined the term, "Supplemental Files." The starting assumption was that the report for the electronic data submission would be made using PDF files. PDF is an acronym for "Portable Document Format." PDF is a file format created by Adobe that is platform independent and enables a user to view and print a file exactly as the author designed it without needing the application or fonts used to create the file.

Each study submitted to the Agency in support of pesticide registration generally consists of a textual summary report with tables of summary data, illustrative figures if appropriate and a number of appendices. The appendices may include a copy of the protocol, analyses of the test substance, detailed data on individual animals or test sites, analytical methods, and analytical data such as chromatograms.

The assumption was that the summary portion of the report (including text and summary tables) would be saved as a PDF file, directly from the word processor used to create the report, and not scanned from a paper copy, with a PDF file created from the image. However, due to the variety of systems that may be used to create tables and appendices of data in the laboratory, it was recognized that some of these might require that PDF files be created from the scanned paper copy. The overall format of the study report is being discussed in a separate chapter entitled Electronic Data Submission: Pilot Efforts in the Office of Pesticide Programs, U.S. EPA, along with results of EPA review of the first pilot electronic data submission.

At the workshop, the two groups came up with two slightly different definitions. Neither group liked the term "supplemental files," and one came up with an alternate term, "Review Aids." At this point, the Agency continues to use the term, "Supplemental Files." We have combined the definitions from the two groups.

"Supplemental Files (Review Aids) are: Any data set needed by the reviewer for additional analysis purposes that cannot be readily extracted from the PDF report, or, additional information in electronic format that would enhance the reviewer's understanding or facilitate presentation of the data."

Therefore, supplemental files are not equivalent to appendices of individual data, although they may sometimes contain the same information. Additionally, supplemental files would not include new data which were not included elsewhere in the study report. They would not include summaries of data from across multiple studies. Supplemental files are not required by the Agency in order to complete a study review because the information or data must be in the paper or PDF copy of the report. Supplemental files are not needed for every type of study, nor for every type of data included in a particular study.

The types of files the groups at the March 2000 workshop agreed would be useful for reviewing studies depended on the type of data being submitted. Some examples of these types of files are listed in Table III.

Table III. Examples of Supplemental Files

- Data tables from mammalian toxicity studies
  - body weights
  - ante or post mortem observations
  - clinical chemistry
  - tumor incidence
- Data tables from residue chemistry and environmental fate studies
  - analytical method validation
  - residues reported by sample
    - crop field trials
    - dislodgeable foliar residues
    - residue dissipation
    - water monitoring
- Chemical structures in ISIS compatible format
- Metabolic pathways (metabolism studies) in ISIS compatible format
- Photographs (e.g., slides, crop production or processing, crop injury)
- Video (e.g., crop production, or food processing procedures)
- Full text (including graphics) of analytical methods or study reports
- Models (Spray Drift, Efficacy, ARTF)
- Dietary exposure input files (including residue distributions)
- PRZM input files

Issues and Concerns

The Agency science reviewers feel that the advantages of having supplemental files are many, that they would primarily increase efficiency and quality of science reviews. The reviewers would have less data entry to do to check the reported results and to perform statistical calculations on the data. They could more easily add text and tables and graphics to their reviews. Chemical structures and metabolic pathways would not need to be redrawn to be inserted into reviews or Agency databases. The reviewers could have on-line access to the actual study data at internal EPA meetings. Not needing to retype data would result in increased accuracy in data analyses and in reviews, more efficient use of review time, and more comprehensive use of the study data. OPP heard some concerns at the Electronic Data Submission Workshop in March 2000, which is discussed in more detail below.

Formatting

The data submitters do not want EPA to implement strict mandatory formatting of supplemental files, and mandatory software requirements. Also, there was concern that OPP may change guidance on supplemental files and require early submitters of supplemental files to resubmit their files in a different format. Strict formatting may force the data submitters to purchase specific brands of software to produce the files desired by EPA. However, OPP is committed to open standards for file submission formats, and has no plans to make the submission of supplemental files a mandatory requirement. OPP plans on being flexible on the formatting of supplemental files.

Over-Analysis

Concern was expressed that, presented with the data electronically, the reviewers may more easily over-analyze the study data. While access to the data electronically will make data analysis easier, the time frames for review of pesticide data are relatively short, not leaving time for over-analysis of the data.

Archiving

Some data submitters were concerned about whether the supplemental files would be archived along with the data, and whether the supplemental files would be subject to Freedom of Information Act (FOIA) requests. OPP expects to maintain the compact disc (CD) provided as the electronic data submission, and archive the supplemental files on CD along with the rest of the electronic data submission. However, the official archive format at the present time is paper, and the supplemental files duplicate material in the paper submission. It is expected that supplemental files can be released upon FOIA requests after the first registration for the pesticide active ingredient, subject to the requirements of FIFRA 10(g), which include an affirmation statement from requestor, and notice to the data owner, excluding any FIFRA CBI.

Certification of Authenticity

Concern was expressed about certification of authenticity, or whether the data submitter would guarantee that the data in the supplemental files was identical to the data in the paper submission. If the supplemental files are created at the same time as the PDF files of the of the study report, there shouldn't be any difficulty in guaranteeing that the data in the supplemental files match the data in the study report. OPP will require that the data submitter sign a Certification with Respect to Data Integrity.

Security

There was also concern about the security of the electronic data submission, including supplemental files, particularly improper access to Confidential Business Information (CBI). The electronic data submission has the same security requirements as a paper submission. The difference is the smaller physical size, resulting in increased portability and ease of copying. OPP science reviewers must be cleared to handle CBI, and the requirements for handling CBI properly are no different for an electronic submission than for a paper submission. At this time, OPP plans to make the electronic data submission available to science reviewers on the Local

Area Network (LAN), but at the present time cannot place CBI on its LAN. OPP expects approval of a security plan that will permit FIFRA CBI on the LAN. An alternative in the meantime would be to provide reviewers with copies of the data on a CD if there was a CBI claim.

Current Activities

OPP has been involved with informal electronic data submissions for several years, predating the recent pilot electronic data submissions. Activities have ranged from informal requests for electronic data from the science reviewer to the formal pilot efforts now ongoing.

At the present time, OPP scientists are evaluating PDF submissions and supplemental files. For a pilot electronic data submission, OPP has discussed the content of supplemental files with the registrant interested in providing an electronic data submission, and has worked on the data fields of interest for several toxicology studies. OPP is evaluating SAS-XPORT (SAS Transport) files as a neutral file format for data tables. Two programs being evaluated by OPP are DBMS Copy and Stat Transfer to transfer the data from the SAS-XPORT format into a format usable by the reviewer.

OPP scientists are identifying the data elements (or fields) of interest for each study type where the reviewers felt that supplemental files would be useful, identifying other information for each study type which would be useful to have as electronic files, and developing guidance for submission of supplemental files. This guidance is not ready to be published, but the scientists will work with the data submitters to ensure that the supplemental files for the particular study will be useful to the science reviewers. Even when guidance becomes available, OPP intends to be flexible about the format of the supplemental files, and will modify the guidance as needed.

So far, the majority of the experience of the OPP science reviewers has been with informal requests for electronic data submissions. These informal submissions laid the groundwork for the more formal pilots. For the informal submissions, the reports generally have been submitted in word processing formats and the data tables in Excel or word processor format. The science reviewers have been enthusiastic, and have felt that reviews were completed much more quickly.

Examples of Experiences with Pilot and Informal Submissions

Residue Chemistry

One electronic data submission has been received for the pilot. The submission included studies for plant metabolism, crop field trials, analytical method, and storage stability. The reports were submitted in PDF, with the appendices to the report submitted as paper copies only. The science reviewer used only the PDF copy of the summary report, which was well formatted with bookmarks, and well documented. A paper copy of the appendices was used in the review. Some text was copied directly from the PDF summary report and attributed to the company. Some data tables were copied from the PDF summary report and reformatted slightly. The reviewer had some difficulty getting one table copied into Excel, and felt that the data tables would have been more useful as supplemental files. This review resulted in a significant time savings, but supplemental files would have saved the reviewer even more time.

A second informal submission is undergoing review in OPP. This is a Task Force submission of monitoring data. The summary report is in Word with appendices on paper only. The data tables are in Excel. The reviewers like the ease of viewing and manipulating the data in the spreadsheet, and expect significant time savings in the review process, although the time savings may be difficult to quantify.

A number of other electronic data submissions have been made informally. These generally consisted of a summary report and data tables in a word processing format. The reviewers felt the time savings was substantial, but noted a difficulty transferring data tables from Word to Word Perfect.

Occupational-Residential

The science reviewers have worked informally with Task Forces on the format for electronic data submissions of large data sets. A data set is under review now. The electronic format facilitated rapid summary of the data in the submissions.

Toxicology

The science reviewers have been working with a pesticide

registrant to identify the data fields needed for several toxicology studies - a chronic feeding/oncogenicity study in rodents, a chronic dog study, a developmental neurotoxicity study, and a multi-generation reproduction study. The studies have not been submitted yet to the Agency for review. Supplemental files using SAS Transport will be included.

In several instances, informal requests to data submitters resulted in submission of supplemental files of tumor data from carcinogenicity studies. This enabled the statistical analyses of these data and resulted in substantial savings in data entry and analysis time.

Other informal experiences included individual science reviewers requesting electronic copies of the submission and data tables. Generally the report and data tables were provided in word processing format. These resulted in significant time savings during the review process. One electronic data submission was made in the CADDY format with no supplemental files. The science reviewers found no time savings with this format.

Environmental Fate and Effects

The experience of the science reviewers has been with informal requests for electronic data submission made directly to the pesticide registrants and disks containing raw data being received along with the paper submissions. Pesticide registrants have been including with their paper submission a diskette with raw data from avian reproduction and terrestrial plant studies for about 5 years. Data for terrestrial field dissipation and water monitoring are often received in spreadsheets. The result was significant time savings and accuracy in analyses of the data.

Future Developments

OPP will continue with pilot electronic data submissions with supplemental files. We will continue to refine the data elements needed for each type of study. We need pilot electronic data submissions with supplemental files to help with this refinement. OPP is willing to receive supplemental files of other types as well, especially chemical structures and metabolic pathways in ISIS compatible format, and photographs and video files.

Watch the EPA Web Site for guidance on submission of supplemental files with Electronic Data Submission, or contact the author for more information. The following web sites may be helpful to the reader for additional information:

- http://www.epa.gov/pesticides
- http://www.epa.gov/oppfead1/edsgoals.htm

Conclusions

In summary, supplemental files received to date have been very useful in the analysis of data, resulting in saving review time. Supplemental files have been most useful for large data sets. EPA has been flexible in the acceptance of supplemental files. Supplemental files have been received in a number of file formats, and all have been useful. Receipt of the study reports electronically is useful as well, saving time in the summarizing of the study. Most of OPP's experience to date has been with supplemental files of data tables, model input parameters, and full text of study reports.

Chapter 26

Current State of Electronic Submissions in Europe

Steven C. Dobson

Pesticides Safety Directorate, Mallard House, Kings Pool, 3 Peasholme Green, York YO8 8QA, United Kingdom

The CADDY (Computer Aided Dossier Design and Supply) system has been developed as a practical, flexible and efficient platform for the electronic submission of regulatory dossiers. The CADDY system is in use within the European Union (EU). This paper provides information on the principles behind the development of CADDY; the benefits of CADDY to both industry and regulators; the current status of CADDY adoption, focussing primarily on EU Member States; and future developments.

Introduction

The driving force for the development of a standardized format for electronic submission of regulatory dossiers for agrochemicals in Europe dates back to 1993. In that year, European Council Directive 91/414/EEC (*1*) came into force and provided a harmonized basis for the regulation of plant protection products across the 15 Member States of the European Union (EU). As part of that legislation, a program for the re-evaluation of all active substances on the European market was established. Several hundred active substances were to be included in this program, which was envisaged to be completed within 10 years. In 1995,the first list of 90 active substances was identified for re-evaluation, and industry was required to submit dossiers for these substances in a standard format. The same format also applied to applications for new active substances.

At the start of the program it became obvious that compiling, transporting and handling these paper-based dossiers imposed a considerable burden, in terms of costs and staff time, on both industry and regulatory authorities. EU dossier submissions contain up to 100,000 pages, the majority of which comprise the individual regulatory studies. In addition, a structured and detailed overview is also supplied by industry (often as word processor/spreadsheet files) as well as additional supporting documentation. Also, in several Member States more than one copy of the full dossier is required, resulting in over 30 copies of the dossier having to be prepared and shipped to support one active substance. Against this background, Member States faced the need to find ever increasing storage space to cope with paper-based dossiers submitted to support both existing and new active substances.

Development of an Electronic Submission Format

To address these difficulties, a working group, formed of delegates from the European Commission, EU Member States, and European industry [represented by the European Crop Protection Association (ECPA)] first met in 1995 to discuss proposals for the development of a standard format for electronic dossier submissions. This group became the EU Member States/ECPA Data Transfer Steering Group (the CADDY group) and the system that they developed became known by the acronym CADDY (Computer Aided Dossier and Data Supply).

By early 1996, the CADDY group had identified an initial strategic goal for the CADDY system which was to facilitate, in a cost-effective manner, using electronic media:

1. the provision of dossiers for plant protection products to regulatory authorities;
2. the long-term archiving of such dossiers, and;
3. the accessibility of information contained in such dossiers.

The CADDY group determined that the system should have a flexible transfer interface that served the individual needs and requirements of end users. In addition, the system should be modular, capable of incorporating and integrating new technology and adapting to changing regulatory needs. The CADDY specifications were built around the following objectives, that:

1. the first release should be very simple and cover only those requirements that were absolutely necessary;
2. the page format (stored dossier pages) and index file format (the indexing system necessary to provide efficient document/information retrieval) should be readable by a wide range of standard applications;
3. the retrieval software should meet the needs of users and allow a CADDY submission to replace a paper copy, if desired;
4. the storage and transfer medium should be CD-ROM, so that a complete dossier could be submitted on 2-3 discs;
5. all the pages of the dossier should be represented as TIFF (Tag Image File Format), which are readable by a wide range of standard imaging applications; and
6. the index information should be represented in a format readable by standard database applications.

From the outset, it was considered that a system that met these objectives would be sufficiently flexible to offer a potential basis for world-wide harmonization of electronic submissions. Thus, in the autumn of 1996, the CADDY Group was expanded to form the Joint Data Transfer Steering Group to include representatives from the US Environmental Protection Agency (EPA), the American Crop Protection Association (ACPA), the Canadian Pest Management Regulatory Agency (PMRA), and Canadian industry representatives. The specifications for CADDY were actively developed by this expanded group, and the requirements of the North American participants were incorporated into both the format and retrieval specifications. Against this background the specification and components of the system were completed. The main components of the system are shown in Table I.

The retrieval software was developed to allow rapid access to the dossier information and to meet the initial needs of users in both industry and Regulatory Agencies. The range of functions available to the regulatory evaluator when working with an electronic submission is an important factor in determining its successful adoption. Therefore, the functions implemented in the CADDY retrieval software provide the evaluator with the ability to:

- display/print dossier details.
- access studies via either a hierarchical and expandable table of contents or a study (report) list.
- bookmark studies.

- annotate studies with private (evaluator only) or public (all users) comments.
- conduct bookmarks and annotation searches.
- select and sort studies by using a comprehensive multi-attribute search tool.
- print study lists and details, print studies and study pages.
- navigate studies using a 'toolbox' (goto next/previous page; first/last page; page orientation/zoom; goto by page number).
- select text/figures from studies and use optical character recognition to export selections to word processor/spreadsheet files.

Table I. Components of the CADDY System

Format specification	Defines the format for compiling a CADDY submission. First version finalized in January 1996. Current version 1.1, finalized September 23, 1997. Freely available.
Compilation software	Compiles a CADDY submission. The development of the software necessary to compile a CADDY-compliant submission, in line with the format specification, is the responsibility of industry/commercial organizations.
Conformity test software	Reports if a dossier on CD-ROM conforms to the CADDY format specification. Current version Revision 18, April 20, 1998. Freely available.
Retrieval software	Allows the submission to be displayed and worked with by Regulatory Agencies and others. First version finalized in February 1997. Current format specification 1.2 (version 1.0) August 1999. Software and support available at nominal cost.
Information brochure	Provides an overview of the CADDY project. Current version dated March 1998.
Application guide	Describes how to deal with CADDY and what to expect from a CADDY submission. Current version dated April 22, 1998 (*2*).

Adoption of CADDY Within Europe and Initial Experiences

By the end of 1997 the first release of the retrieval software completed the components of the CADDY system and allowed the initial strategic goal of the CADDY group to be achieved, which was to provide a practical and workable electronic submission system. To date, 53 organizations worldwide (43 in

Europe, 8 in North America and 1 each in Japan and India) have purchased licenses for the retrieval software and help desk support, covering a total of 563 licenses. Organizations range from regulatory authorities, research based manufacturers, generic manufacturers to consultancy/contract houses. The system has also been demonstrated to a wide range of interested organizations involved in pesticide regulation, including authorities considering biocides in the EU; the Food and Agriculture Organization in Rome; the World Health Organization in Geneva; regulatory authorities in Central and Eastern Europe; as well as the Organization for Economic Co-operation and Development's Working Group on Pesticides.

CADDY has been adopted as the standard for electronic dossier submissions for plant protection products within the European Union. For example, since the release of the retrieval software in 1997, the UK has received a total 15 CADDY submissions, with similar numbers having been received by other Member States. The system has led to improved efficiency in dossier compilation by industry and significant savings in handling and shipping costs. On this basis companies have been quick to realize the benefits of electronic dossier assembly. Production of electronic submissions in CADDY format from in-house document management systems, such as Documentum, has reduced the time to assemble dossiers and associated costs. Novartis estimated that a recent EU CADDY submission, comprising 8 paper copies and 32 CADDY dossier copies, represented a saving of over $100,000 compared with paper-based dossiers alone.

For regulatory authorities in Europe, the adoption of CADDY has provided a means of alleviating the burden of handling and storing paper based submissions. Initial experience with the CADDY system has involved its use as an aid to dossier completeness checking and as a tool for rapid access to underlying studies in support of decision making. So far the use of CADDY based submissions for evaluation has focussed on discrete parts of the regulatory submission, for example, physical chemical properties and methods of analysis.

While initial feed back from users in Regulatory Agencies has been positive, CADDY, like other electronic submission methods, is unlikely to replace fully the need for paper submissions in the medium-term. Important ergonomic issues remain to be resolved before full evaluations of active substance dossiers routinely will be conducted electronically, and a true paperless submission process is achieved. In particular, a general handicap of electronic submissions is the readability of the text on the screens. The resolution and size of screens remains a limiting factor for the computer based evaluation process. This is not a CADDY specific issue but is applicable to all types of electronic submission, as today's screen technology is still not comparable to paper in terms of readability or ease of use for prolonged periods.

As a first step toward this goal of a true paperless submission, the potential exists to use CADDY as a replacement for paper when evaluating limited data packages, such as those supporting product registrations.

Future Development of CADDY

In 1999, to address the needs of regulators and industry to fully utilize CADDY as an evaluation tool, the CADDY Group revised the strategic goal to include 'the examination and assessment of dossiers by regulatory authorities'. This acknowledged the desire of the regulators and industry to enhance and develop the potential of CADDY as an evaluation tool. As part of this process, the retrieval software was further enhanced with the release of an 32 bit version with improved functionality and network support.

In order to assess the requirements for the further development of the retrieval software, industry members of the CADDY group examined the current use of CADDY in 6 European regulatory authorities early in 2000. The analysis found that overall there was significant interest in CADDY but only evaluators in the larger authorities had actual experience with the software. While the current CADDY functionality was judged by evaluators in the 6 countries to be sufficient for their immediate needs, further improvements were identified.

Therefore, to increase the use and acceptance of the CADDY retrieval software, the following measures have been identified for action:

1. Improve functionality - in the next stage of development the addition of hyperlinks between summaries and reports, improved copy, pasting of tabular data, and full text search will be considered. In late 1999, the Steering Group adopted a future vision for CADDY that included additional elements coded using XML (eXtensible Markup Language) to enhance functionality with structured data and provide a basis for long-term development;
2. Improve online help and establish an e-mail hotline;
3. Provide updated documentation and training material with every new version;
4. Improve user support and decrease response time with in-house or offsite training;
5. Visit authorities regularly to promote and assess the adoption of CADDY;
6. Update the CADDY website with new areas like discussion groups and FAQ's; and

7. Improve industry quality control systems to ensure high quality CADDY dossiers are submitted that fully comply with the format specification.

Conclusion

The adoption of CADDY within the EU has provided a sound basis for the transfer of regulatory dossiers between industry and regulators. CADDY has been developed from inception to a workable system in 5 years and represents a significant collaborative effort between industry and regulators drawn from North America as well as Europe. This initial achievement, has resulted in substantial reductions in compiling, transporting and handling costs for EU dossiers and has provided a basis for rapid dossier access. CADDY is now being further developed to build upon and enhance its capabilities for use as an evaluation tool. Further information on CADDY is available on the ECPA website (www.ecpa.be).

References

1. Council Directive 91/414/EEC of 15 July 1991 concerning the placing of plant
protection products on the market, O.J. No L 230, 19.8.1991.
2. Guidelines for preparation of dossiers by applicants. Doc 1663/VI/94 Rev 8 of 22.4.1998.

Chapter 27

Pest Management Regulatory Agency Experiences: Electronic Submissions

Carmen Krogh

Health Canada, Pest Management Regulatory Agency, Room D746, Sir Charles Tupper Building, 2720 Riverside Drive, Ottawa, Ontario K1A 0K9, Canada

The Pest Management Regulatory Agency (PMRA) was created in 1995. The PMRA regulates all pesticides, (ie., products designed to manage, destroy, attract, or repel pests), that are used, sold, or imported into Canada.. It conducts science-based health risk, environmental risk, and value (including efficacy) assessments of each pesticide before determining if a pesticide product should be approved for use in Canada. By seeking to minimize the risks associated with pesticides, the PMRA helps protect human health, safety, and the environment.

The PMRA is committed to improving processes and to reducing costs associated with the review of submissions. The approach to meeting this commitment includes the application of process change management activities, international harmonization, and achieving an electronic capability through the application of technological solutions. These activities have formed the basis of the PMRA electronic submissions and review project. The project is international in nature with linkages established with the U.S. Environmental Protection Agency (EPA) - Office of Pesticide Programs (OPP), the European Union (EU), the Organization for Economic Co-operation and Development (OECD), and the pesticide industry in Canada, the U.S., Australia, and Europe. The PMRA participates on the North American Free Trade Agreement (NAFTA) Technical Working Group (TWG),Regulatory Capacity Building sub-committee. Under this working group, PMRA and EPA are working with the North American pesticide industry to pilot joint projects in the electronic submission and review environment. This involves moving from paper-based processes to electronic ones.

Electronic Harmonization

Currently, there are a number of harmonization activities in progress. Most of these involve harmonization of data and dossier formats and tend to be paper based. Harmonization efforts on work sharing are also currently underway among international regulatory authorities. Work sharing is facilitated by common formats for data submissions and review as well as electronic tools to make the submission assembly and review a more efficient and effective process.

The benefits associated with an international harmonization approach include increased predictability and consistency of review time, a benefit to both industry and government. As well, harmonization of common formats allows evaluators to focus on the science of the reviews where less time is spent looking for information or reformatting information. With the use of electronic tools, evaluators can reuse information through 'copy and paste' techniques.

Submissions Process

The submission process has three broad components and there has been positive progress in the first two:

1. Electronic assembly (involves the assembly of the electronic submission in a manner that evaluators can use);
2. Electronic evaluation (involves using the electronic submission for an electronic, desktop-based evaluation); and,
3. Electronic archiving (is needed by both industry and regulators).

Joint Industry-PMRA Pilot

PMRA and Bayer, Inc., collaborated on an electronic submissions pilot project. Its purpose was to evaluate an electronic evaluation capability and to prepare for the new way of doing business. The pilot format was designed to support the efficient electronic assembly and review of a submission. Required elements included the ability to deliver, store, update and retrieve information, and provide the PMRA evaluators with improved processes and functionality at the desktop.

Bayer provided 4 submission formats: the Computer Aided Dossier, Delivery and Supply (CADDY) specification, the Portable Document Format (PDF) format, the PDF viewed in a web-browser, and paper. These formats were compared in a methodical and unbiased manner. It was determined that the PDF format gave evaluators a 23% gain in efficiency as compared to paper. Results clearly demonstrated that the PDF format provided evaluators with sufficient desktop functionality and improved efficiency.

While CADDY is a useful tool for European industry and regulatory authorities, it was not embraced by PMRA evaluators because of its limitations as an aid to study review by regulatory staff.

Electronic Submission Success Factors

There are a number of factors that should be taken into consideration when developing electronic solutions. To be successful, the electronic solutions must support evaluator desktop needs and must utilize open standards / web-based tools to provide a neutral non proprietary approach that allows sharing of information internationally. The system must have the capacity to evolve, which dictates close co-operation between industry and regulators. Participants must be ready and prepared to deal with the change in data management requirements as well as work flow processes.

Guidance Available

The PMRA has drafted 3 guidance documents to assist with the electronic submission process. These have been extensively commented on by industry and users in North America.

1. Guidance to Registrants for Preparing Electronic Submissions; Part II: Guidance for Industry During Pilot Stage. This document provides guidance to registrants on assembling an electronic submission.
2. Guidance to Registrants for Preparing Electronic Submissions; Part III: Guidance of Evaluator Functional Requirements for Electronic Evaluation. This document describes evaluator needs for e-review.
3. Guidance to Registrants for Preparing Electronic Submissions; Part IV:

Guidance on Preparation of Documents for Electronic Exchange. This provides guidance on how to create a document to minimize conversion issues between proprietary software such as Microsoft Word and Corel WordPerfect.

They are available on the PMRA web site: http:/www.hc-sc.gc.ca/pmra-arla

Evaluator Needs

The PMRA evaluators determined their needs for conducting an efficient electronic review:

1. Navigation: ease in using bookmarks and links accessed by point and click.
2. Document Viewing/Printing: high quality, viewable on the screen, and (if needed) easy to print.
3. Document Annotation: must be able to add reviewer annotations.
4. Data Manipulation: ability to manipulate data using spreadsheet or other analytical methods.
5. Report Generation: ability to re-use information through copy and paste functionality.
6. Ergonomics: comfortable screen size, PC, and desktop design.
7. Links to other files: includes links to supplemental files such as histograms, video.

Live Demonstration

A joint demonstration of the electronic submission was presented jointly by representatives from the PMRA and Bayer, Inc.

Summary

While the PDF format is not perfect, it meets the requirements for a 'neutral' format which is portable across the international pesticide community. The PDF based pilots are demonstrating positive results.

Outcome expected from future pilots is establishment of an electronic submission formatting standard that strikes a good balance in meeting efficiency gains for both registrants and reviewers. The standard must be cost effective and easy for the wide range of North American registrants to implement. It must provide reviewers with easily learned functionality that makes their work more efficient and effective.

Chapter 28

Electronic Data Submissions for the Environmental Protection Agency and Pest Management Regulatory Agency: Zoxamide Fungicide

Janet Ollinger[1], Paul H. Reibach[1], and Scott Swidersky[2]

[1]Rohm and Haas Company, 100 Independence Mall West, Philadelphia, PA 19106
[2]Quality Associates Inc., Suite 102, 9017 Red Branch Road, Columbia, MD 21045

When a manufacturer petitions a regulatory agency to register a new pesticide, multiple copies of over 100 individual studies are submitted to support the petition. Historically, paper copies have been submitted and required. Submission and review of electronic copies is still a very new and evolving procedure. For registration of a new fungicide, called zoxamide, electronic copies of a text and tables of residue studies were submitted to EPA as part of a pilot program in addition to the paper copies. Additionally, full copies of zoxamide environmental fate studies were submitted electronically to PMRA.

Registration of Zoxamide Fungicide

The Rohm and Haas Company provided electronic data submissions to both the United States Environmental Protection Agency (EPA), to Canada's Pest Management Regulatory Agency (PMRA), and to Mexico's CICOPLAFEST to support the registration of a new fungicide, called zoxamide. Rohm and Haas applied for the registration of zoxamide in December 1998 for control of late blight on potatoes and downy mildew on grapes. Zoxamide is an important product from a regulatory viewpoint because the submission had two regulatory "firsts":

- This will be the first joint tri-lateral NAFTA review and registration decision of a pesticide by EPA, Canada, and Mexico. Previous joint reviews of pesticides were between EPA and Canada;
- The submission included an international Organization for Economic Cooperation and Development (OECD) summary dossier, the first OECD dossier provided to EPA, Canada, and Mexico.

Zoxamide will also be an important product for the growers, especially the potato growers who need as many products as possible to control late blight, a devastating disease. Zoxamide has a very favorable toxicology and environmental fate profile. It is not acutely toxic, not genotoxic in mammalian systems, not neurotoxic, not oncogenic, and not developmentally or reproductively toxic. It degrades rapidly in soil and water, is not mobile in soil, and will not contaminate groundwater. Furthermore, it does not have adverse risks to birds, earthworms, bees, or other beneficial organisms.

The initial submission of all studies to support the zoxamide registration with EPA/PMRA/CICOPLAFEST was a paper copy because Rohm and Haas had not planned to make an electronic submission. Thus, the dossier needed to be retrofitted for the electronic copies. Rohm and Haas appreciates the guidance given by both PMRA and EPA in providing the electronic submissions.

Rohm and Haas provided two separate and distinct forms of electronic submissions to support Zoxamide's registration:

1. EPA pilot program: electronic copies of the text and tables from residue studies were provided;
2. Canadian program: entire reports of environmental fate studies including text, tables, graphs, etc., were submitted electronically following the Canadian draft guidance for electronic submissions.

Components of a Pesticide Registration Submission

The objective of the electronic submissions was to facilitate the regulatory review and to increase review efficiency. The registration is still pending, so an assessment of this objective is not yet possible.

Types of Studies Included With a Registration

Before discussing the electronic submissions, it is worthwhile to review the types of studies that are included in the dossier. These include physicochemical properties (hydrolysis, water solubility, octanol/water partition coefficient, etc.), metabolism (soil, plant, animal), environmental fate (adsorption/desorption, leaching, water sediment, field dissipation, etc.), residue studies (field trials, methods, residue analysis, processing, etc.), ecotoxicology (birds, fish, algae, etc.), toxicology (acute, chronic, oncogenicity, genotoxicity, developmental toxicity, etc.), and risk assessments. Each study fulfills a particular requirement. The requirements are called a guideline number at EPA, a data code or DACO number in Canada, or an Annex Point Number under the European Union (EU) and the OECD system. For example, the hydrolysis study is OECD Annex Point Number IIA 2.9.1 in the OECD format and 2.9.1 in the EU, or 835.2120 in EPA's OPPTS system and 161-1 in EPA's OPP system, and 8.2.3.2 in Canada's DACO system (Table I).

Table I. Sample Guideline Numbers

Guideline	Hydrolysis	Aged Column Leach
OECD	IIA 2.9.1	IIA 7.4.5
EU Annex IIA	2.9.1	7.1.3.2
EPA OPPTS	835.2120	835.1240
EPA OPP	161-1	163-1
Canada DACO	8.2.3.2	8.2.4.3

Different Forms of Information in a Report

A report typically has information in many different forms, including:

- Text
  - Typically MS Word Document
- Tables
  - Excel Spreadsheets
  - Crop residue data tables in Word
- Chromatographic Data
  - HPLC and GC chromatograms
  - LC and GC mass spectra
  - Run Sheets
  - Data Reprocessing
- Other Information
  - Photographs
  - TLC Images
  - Contract lab reports

Zoxamide Electronic Submissions to EPA and PMRA

For the Zoxamide registration, EPA and PMRA divided the study reviews. EPA reviewed residue and processing studies, among others, and PMRA reviewed the environmental fate studies, among others. Because the reviews were shared between the two regulatory agencies, two separate and distinct electronic submissions were prepared:

1) A submission of electronic text and tables from Zoxamide residue reports to EPA;
2) A submission of the complete and full environmental fate studies to PMRA.

EPA Electronic Submission–Pilot Program

A residue submission was made to EPA as part of their electronic pilot program. It was possible to provide electronic copies of the residue report text and tables because the reports were originally prepared with text in Microsoft Word® and data tables in Word® or Excel® formats. These parts of the report would aid the reviewer because data tables could be incorporated into the review and would not need to be re-typed. However, it was not possible to provide the entire report because the chromatograms and other raw data were not available electronically. Thus, the electronic portions available, including text and tables, were converted to PDF format using the ADOBE Acrobat® software, and then sent to EPA. This was a very easy operation to carry out.

PMRA Electronic Submission

The zoxamide environmental fate studies were submitted electronically for the PMRA submission because PMRA was reviewing this portion of the submission. The studies were formatted using PMRA's Electronic Dossier Delivery and Evaluation (EDDE) pilot guidelines, which are available at PMRA's web site on the World Wide Web at http://www.hc-sc.gc.ca/pmra.arla.

As noted previously, the original zoxamide submission was a paper copy. For the environmental fate studies, the report and data tables were in Word and Excel formats; however, the majority of the report was comprised of chromatograms and other data that were not available electronically. We decided that it would be easiest to start from the paper copy of the entire report. Thus, the entire reports, including text, tables, chromatograms, and other pertinent raw data, were scanned and prepared in electronic format.

The submission was prepared in a portable document format (.pdf) {image + hidden text, to be exact}. The reasons for supporting this format were to maintain the original document as a graphic file, while providing reviewers with full text search capabilities, and to provide the regulatory agencies with a non-platform specific format. Also, with the .pdf format, tables and text can be copied directly from an image file to another source document, without requiring data-entry.

In order to track each of the reports electronically, an index describing the profiles of each of the studies using MS Access Database was created. This MS Access Database (Figure 1) had elements required by PMRA's EDDE, such as the File Name and Key Words, and other information that was not required by PMRA. For example, a line for the EPA Master Record Identification Number (MRID), was included because both the original paper submission and the electronic submissions were sent to EPA, PMRA, and to Mexico. For that reason, the EPA Guideline Number was also included in addition to PMRA's DACO number. Additional information was also populated in the .pdf (document information fields) to provide "catalog search" functionality for searching through multiple documents.

As shown in Figure 1, the information in this database included:

- Title
- Author
- Chemical
- DACO Number
- Letter of Designation
- Study Type
- Animal Species
- US Guideline
- Study Number
- MRID
- File Name
- Key Words

After entry into the MS Access database, the reports were scanned as .TIFF images. The concatenation process from the MS Access database to produce the file name was used as the index for each of the documents. Once each of the documents was scanned and assigned its appropriate file name, it was converted from a .TIFF file to pdf (image + hidden text).

Utilizing the data produced in the MS Access file, a relation was created by file name and the remainder of the data to populate the assigned document information fields. Bookmarks and Hyperlinks were then created manually, because it was not possible to do this electronically, although several keystroke methods ensured that the process was efficient. The original Table of Contents for the reports was used to determine the significance of the Bookmarks.

A Visual Basic Program was developed to create the directory structure and naming convention to facilitate the location of the data files. The program was designed to create organized DACO categories, whereby the report that

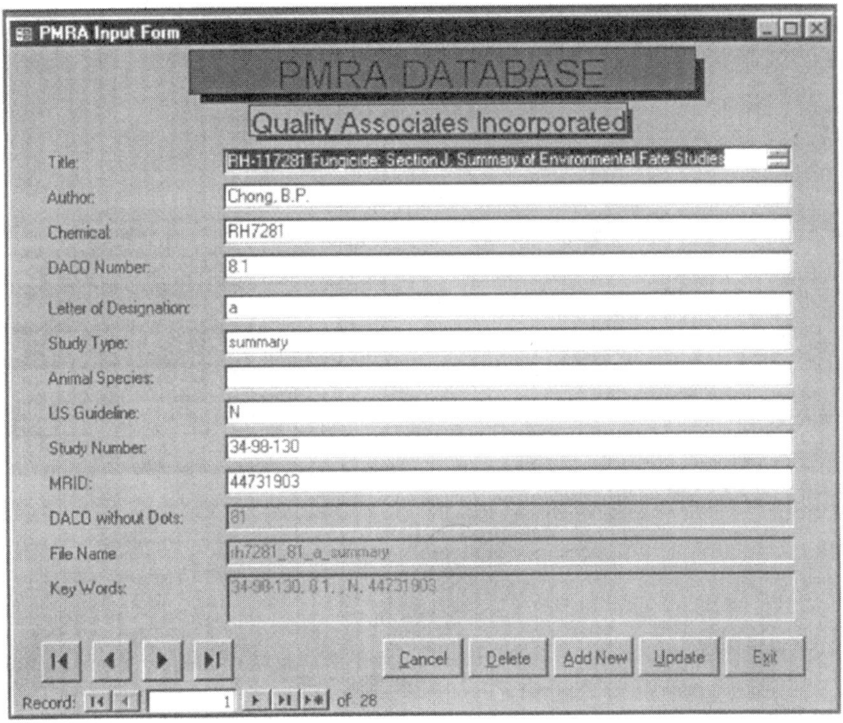

Figure 1. File Directory Structure.

corresponded to a specific DACO requirement was entered into a folder with the specific DACO number. The directory structure is shown in Figure 2.

With this format, a reviewer can simply locate the appropriate DACO folder, click on the folder, and find the report. Bookmarks and Hyperlinks help the reviewer find specific information.

Another item that was created for the zoxamide PMRA electronic submission was the "Main Page". This page is not required by the EDDE Guidelines, but its creation greatly facilitates the use of the database. An example of the Main Page is shown in Table II. With all of these features, it is easy to find and search specific documents for the information desired.

With the use of the Main Page, clicking on the DACO number or MRID number easily accesses the reports. Because the submission was a multi-agency submission, the "Main Page" enabled the different agencies to relate to the appropriate attributes.

Benefits of a PDF Submission

The benefits to the reviewer are that:

1) The reviewer can find and retrieve reports efficiently with hyperlinks and bookmarks,
2) The reviewer can navigate for keywords and phrases,
3) The reviewer can incorporate electronic data/tables without recreating them, and
4) The electronic submission can potentially reduce document storage space.

Benefits to the registrant are:

1) The registrant can more easily share electronic documents with multiple countries,
2) There are efficiency gains because fewer copies are required, and
3) Potentially shorter registration times due to increased efficiency with reviews.

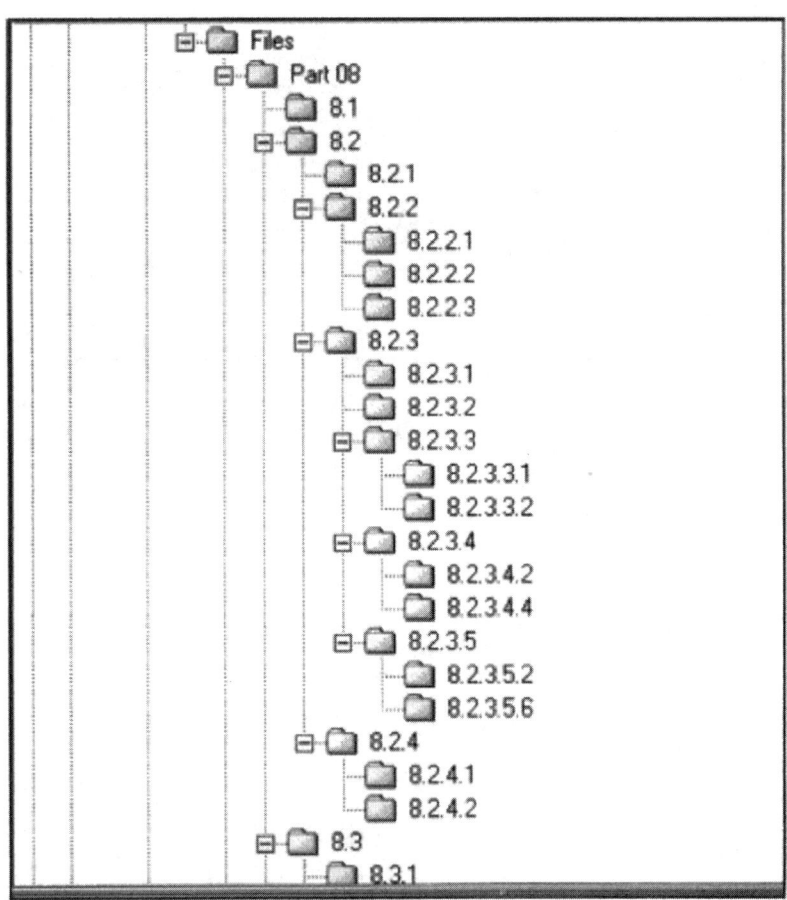

Figure 2. Database Example.

Table II. Main Page Summary

Chemical RH7281 Zoxium 80W					
"Clicking on the blue description for each report allows access to the documents"					
DACO NUMBER	STUDY TITLE	STUDY TYPE	STUDY NO.	US GUIDELINE	MRID
8.2.3.5.2	14C-RH-117281: Degradation and Metabolism in Aquatic Systems	aerobic aquatic metabolism	–		–
8.1	RH-117281 Fungicide: Section J, Summary of Environmental Fate Studies	summary	–	N	–
8.2.1	RH-117281 Fungicide: Section J, Summary of Environmental Fate Studies	summary	–	N	–
8.3.1	RH-117281 Fungicide: Section J, Summary of Environmental Fate Studies	summary	–		–
8.2.4.1	RH-117281 Fungicide: Section J, Summary of Environmental Fate Studies	summary	–	N	–
8.2.3.1	RH-117281 Fungicide: Section J, Summary of Environmental Fate Studies	summary	–	N	–

Chapter 29

The Legal and Policy Framework for Electronic Reporting of Environmental Compliance Reports: Challenges of E-Government: Maintaining Effective Stewardship of the Environment

M. Evi Huffer

Office of Environmental Information, U.S. Environmental Protection Agency, Mail Stop 2823, 1200 Pennsylvania Avenue, N. W., Washington, DC 20460

USEPA has a long commitment to E-Government, having recognized the many benefits of e-commerce, particularly the opportunities to improve Agency business processes and management of environmental data. This paper discusses EPA's progress in establishing the legal framework to introduce electronic reporting/record-keeping (ERR) for environmental compliance documents. It focuses on some of the unique legal challenges faced by EPA in implementing ERR and provides an overview of the Agency's draft proposed electronic reporting and record-keeping rule.

---

[1]The views expressed in this paper are the author's, and do not necessarily reflect the official position of the United States Environmental Protection Agency.

In recent years, we have seen the dawn of a new age of Electronic Government (E-Gov) and with it new opportunities and challenges. While to a number of federal agencies the advent of E-Gov may be taking them down completely new roads, the United States Environmental Protection Agency (EPA) has a long history of electronic reporting activities and working to address issues related to electronic government (1). EPA is building the technological and institutional infrastructure to make E-Gov a reality within the Agency.

A major component of the institutional infrastructure is developing the legal framework for E-Gov. To this end, EPA has worked diligently assessing the requirements of all its stakeholders including: industry, state and local governments, environmental groups, other non-governmental organizations, as well as the general public. Over the years, EPA has also held various forums that brought together government, industry and legal authorities who have extensive knowledge and understanding of the legal and regulatory processes (2). This paper will provide an overview of some of the opportunities and challenges EPA faces developing the legal framework for its electronic reporting/record-keeping program for environmental compliance regulations. It will also discuss the Agency's proposal for electronic reporting and record-keeping which will establish that legal framework within EPA (3). Addressing the legal issues and establishing the legal framework are perhaps the most challenging in moving toward E-Gov for a regulatory agency like EPA.

Mandates

What prompted EPA to embark on Electronic reporting and what are the forces driving EPA toward E-Gov today? Simply stated, EPA began its electronic reporting initiative because it is good business and -- more recently -- it has a clear federal mandate. There have been a number of forces in the last decade, both internal and external, pushing the Agency toward electronic commerce. One of the early drivers behind EPA's electronic reporting program was the Paperwork Reduction Act (PRA), implemented by the Office of Management and Budget (4). The PRA encouraged federal regulatory agencies, like EPA, to reduce public burden and costs associated with information collection (5). It was clear, even then, that significant other benefits existed for those who automated reporting. Benefits such as improved data quality, ability to collect real-time data, and the potential to streamline EPA's business processes. All of these forces increased pressure on Agency's like EPA to not only reduce costs and burden on states and the public, but to streamline regulatory and information management processes as well.

The Clinton-Gore Administration's vision of a Federal Government conducting business with the public electronically, is articulated in such documents as former Vice President Gore's *Access America* (6). This E-Gov vision was a driving force in the 1990s for a number of internal EPA initiatives such as the Reinventing Environmental Information (REI) Action Plan (7). The goals behind such initiatives were to reduce burden and costs and improve EPA customer service. EPA's commitment to E-Gov was further bolstered by the creation of the new Office of Environmental Information (OEI), which has been operational since the fall of 1999. OEI has been given responsibility for stewardship of the Agency's information management, policy and technology. It also leads the Agency in promoting and fostering electronic reporting and record-keeping for compliance reporting; as well as building the support infrastructure within EPA. As stated on OEI's web site, its role is to "ensure that EPA collects high quality environmental information and makes it available to the American public. We provide guidance to assist the agency about the way we collect, manage, analyze and provide access to environmental information. By fulfilling these activities we expect that the public and policymakers can make informed decisions". EPA's E-Gov program goals include to not only reduce costs and burden on the regulated community and improve the quality and efficiency of data, but also to establish an integrated electronic system for collecting and managing environmental data and information.

In 1998, the Federal Government received it's first legislative mandate for E-Gov: the Government Paperwork Elimination Act (GPEA) (8). GPEA requires federal agencies to provide regulated companies the option of reporting or keeping records electronically, including the use of electronic signatures, by October 2003 (9). GPEA is significant for it is the first legal step toward realization of E-Gov and marks the first time electronic signatures are given legal equivalency with the traditional "wet-ink-on-paper" signatures in such a statute.

A more recent legislative mandate, the Electronic Signatures in Global and National Commerce Act (E-Sign) of 2000, was enacted to eliminate the legal barriers to commercial, consumer, and business transactions affecting interstate and foreign commerce (10). E-Sign provides that transactions cannot be denied legal effect solely because an electronic document or signature was used in its formation. While the E-Sign legislation was primarily intended to apply to commercial business transactions and is interpreted to exclude uniquely governmental transactions -- those that related principally to the conduct of government business. Nonetheless, it can include some transactions regulated by federal and State agencies, as some records may have both a government and a commercial or business purpose (*11*).

Both the GPEA and E-Sign legislation have increased the urgency for federal and state agencies to adopt electronic reporting and record-keeping approaches, including electronic signatures, for their regulatory programs. Fortunately, EPA is ahead of most other agencies and has been working on electronic reporting of environmental compliance data since 1990. With the advent of the new mandates and the creation of the new OEI, the move to E-Gov has gained center stage and constitutes one of the Agency's most significant programs.

There remains, however, a constant mandate that also drives EPA's E-Gov program -- as well as all Agency programs – and that is the mission of the Agency itself: to protect human health and safeguard the environment. To that end, EPA must ensure full compliance with the environmental laws in place. As stated in the September 2000 *EPA Strategic Plan*, much of the success of our nation's environmental record over the last 30 years has been attributed "to a strong set of environmental laws and an expectation of compliance with those laws". Further, an aggressive enforcement program is seen not only as ensuring compliance but also provides fair competition "in the marketplace by ensuring that noncomplying facilities do not gain an unfair competitive advantage" (*12*). Developing and implementing regulations to protect human health and the environment is what EPA does as an agency, and part of EPA's mission in moving to E-Gov is ensuring that those regulations remain credible and enforceable.

EPA's general strategy for electronic reporting/record-keeping of compliance information consists of defining an agency-wide approach which offers regulated entities (both States and companies) consistent, predictable ways to do business with EPA. This approach will simplify and standardize management processes, offer economies of scale to EPA's program offices, and increase the ability to effectively manage, distribute and integrate data. It's a two-pronged strategy: designed to build the internal systems infrastructure -- the Central Data Exchange (CDX) -- and the legal infrastructure -- the Cross-Media Electronic Reporting and Record-keeping Rule (CROMERRR). While establishing the legal framework through CROMERRR is the focus of this paper, CDX represents a significant development in how the Agency collects its data from the regulated community and provides opportunities for the agency to rationalize its information management processes. CDX will serve as EPA's primary gateway for electronic documents received by the Agency. The intent is to eventually provide -- to the extent possible

-- a single portal for the regulated community to exchange electronic documents with EPA. States will also have the option of using CDX as a gateway for electronic environmental submissions from their regulated community, offering them a cost-effective alternative to building their own individual systems.

EPA has many reasons to collect data. These include: decision-making, planning, trend analysis, performance monitoring/measurement, and enforcement. Given the public's strong interest in maintaining effective environmental standards, credible data are key to identifying problems and addressing them. While enforcement is one of the many uses of data, "it is one that requires the highest levels of credibility ....(and) enforcement actions are the strongest possible test of data credibility. An electronic reporting program providing data that can meet enforcement needs will produce data that agencies can use with confidence for any purpose. The regulated community is then assured that its data are being handled responsibly" (*13*).

Compliance with the nation's environmental protection program is largely based on self-reporting by industry. A regulatory agency like EPA relies heavily on the deterrent effect of criminal prosecution as the primary means for complying with environmental laws. Electronic filings, like their paper counterparts, may serve as evidence in a civil or criminal proceedings. A concern within the enforcement community regarding electronic filings of compliance data with electronic signatures is the lack of actual courtroom experience with such filings. Behind paper-based, wet-ink signatures, a body of experience has developed over the years to analyze handwritten signatures, detect forgeries and alterations to documents. As a result of this considerable experience, there is a significant body of case law regarding authentication, data integrity, and non-repudiation in handling environmental compliance reports with handwritten signatures. However, for electronic filings, such case law is still largely being developed. Further, when electronic documents are used as evidence in proceedings they must first be admissible in a court of law as evidence, and they must also be 'persuasive". While the GPEA and E-Sign legislation prohibit electronic documents from being excluded as evidence solely because they are electronic, the laws do not ensure that juries will find the evidence persuasive.

As the Department of Justice states, in order to protect the government's interests, agency documents need to be available, reliable, and persuasive and "Electronic processes sufficient to protect an agency's position in court should also be able to address any legal responsibilities to these other audiences just as well-designed paper processes" (*14*). Thus, a key component in designing any electronic reporting or electronic record-retention system is to create electronic systems which produce credible data that when used as evidence is both admissible and persuasive.

Legal Framework

Addressing legal issues raised by moving to electronic processes is critical to developing the legal framework and minimizing the legal risks that may compromise EPA's mission, and is an integral part of implementing E-Gov. Such a framework identifies and removes legal obstacles to electronic reporting/record-keeping and ensures the legal validity and enforceability of electronic approaches.

Through its research and analysis in electronic reporting/record-keeping for its regulatory programs, EPA has identified three features that a legally valid regulatory compliance system needs to provide to ensure the enforceability of electronic regulatory compliance programs. They are authentication, document integrity, and nonrepudiation. These three legal/security concepts need to be addressed in developing the legal framework, whether it is paper-or electronic-based.

Authentication refers to the ability to establish that the originator of a transmission is the individual or organization it purports to be. For a document to be authentic, it must be established: who sent the report, when the report was sent, and when the report was received. Proofing the authenticity of an electronic document is generally based on three methods: "something you know", "something you have", and "something you are". Authentication technologies based on "something you know" include passwords and personal identification numbers (PINs). The "something you have" method relies on the originator possesses a specific piece of property such as an encryption token (i.e., smart card) or a digital signature/certificate. The third method, "something you are", relies on biometrics such as an individual's unique characteristics in voice patterns, fingerprints, handwriting attributes. Depending on the level of authentication required, one or more of these methods may be used.

Document integrity refers to the ability to show that the data received are the same data that were submitted and that the data have not been altered in transmission, storage or retrieval. For document integrity, the transmission process must be ensured, and a permanent record of the transmission must be created. Proper record retention procedures and archiving are important components of ensuring the overall integrity of documents.

The concept of 'non-repudiation', while not a formal legal term, is commonly used within the electronic commerce community, and refers to the ability to prove to a neutral third party that the individual who originated the transmission intended to be bound to the content and context (the substance) of the electronic transmission. The 'intent to be bound', when coupled with authentication and document integrity, establishes nonrepudiation. Environmental compliance reports often require officials to sign the reports and attest to its truthfulness, completeness and accuracy. The concept of "non-repudiation" is critical to such documents as it reduces the ability of the originator to disavow responsibility for the document in question.

The Agency has drafted a proposed rule which it believes ensures the authenticity, integrity, accessibility, and non-repudiation of electronic documents. One that will provide electronic documents the same legal and evidentiary force as their paper counterparts. This proposal reflects extensive consultation with EPA's diverse community of stakeholders -- industry, states, tribes, local governments, non-governmental organizations, and other federal agencies; and through a series of state conferences and two EPA public information meetings last year to discuss the proposed rule's approach and CDX.

CROMERRR Proposal

The proposed rule's goals are straightforward but challenging: to provide the regulated community with the option of submitting electronic reports and maintaining electronic records, including electronic signatures, in lieu of paper reports/records and wet-ink signatures; while ensuring that those electronic reports and records submitted and maintained by the regulated community are reliable and trustworthy and available to EPA and state environmental agencies as required by regulation. The multiple objectives include: reducing both costs and burden for regulated companies, allowing flexibility for various approaches, and providing freedom to adopt new technologies as they became available.

Generally, the proposed CROMERRR establishes the legal framework. It removes the existing regulatory barriers to electronic reporting and electronic record-keeping, such as the requirements for "paper" based reports, signatures, and records embedded throughout EPA's current regulations. The proposal provides for compliance reports to be submitted and/or records to be maintained electronically, in lieu of paper, so long as the electronic reporting or record-keeping satisfies the requirements of the rule. Its approach is to identify performance-based criteria that -- to the extent possible -- ensure integrity, authenticity and non-repudiation of electronic reports and records; specifying "technology-neutral" criteria for acceptable electronic reporting and record retention systems. CROMERRR establishes standards for electronic reports and records affected by both the GPEA and E-Sign legislation. The rule's scope is intended to cover all EPA environmental compliance programs, both reporting and record-keeping requirements, and include EPA delegated state programs (15). The rule does not stipulate technology; nor does the rule promulgate any new environmental regulatory requirements.

Since many of EPA's programs are delegated to state agencies, the proposed rule contains provisions for delegated state programs (16). Early in the process, before drafting the proposal, EPA worked extensively with the states through such forums as the National Governor's Association (NGA) to identify state business needs and determine the impact an EPA electronic reporting/record-keeping rule on delegated state programs. For federal programs which EPA has delegated to authorized/approved states, the rule contains provisions for approving state electronic reporting systems and record-keeping programs for implementing federally delegated programs. Basically, the proposed rule sets criteria for approval of delegated state electronic reporting and record-keeping programs when such approval is required as determined by existing state primacy regulations (17).

For submitting electronic compliance reports for a federal reporting requirement, the rule proposes that electronic documents submitted to EPA or delegated state agencies satisfy two requirements. First, electronic reports must be submitted to an EPA designated electronic document receiving system or to an approved state electronic document receiving system. Second, for electronic reports that require signatures, the document must be signed with an electronic signature that can be validated using the appropriate EPA or approved state electronic document receiving system. Rather than specifying complex procedural and technological requirements for companies, the proposed approach requires submitters to use specified EPA and state systems. The proposal then sets general requirements in the form of performance-based criteria for government systems receiving electronically signed reports from regulated entities. The general areas addressed by the proposed performance criteria for government systems include system security, electronic signature method, submitter registration process, electronic signature/certification scenario, transaction record, and system archives.

For electronic record-keeping, the rule proposes some basic provisions in the form of performance-based criteria which regulated entities' electronic record-retention systems must satisfy. The proposed CROMERRR provides that electronic records required to be maintained under EPA programs satisfy federal record-keeping requirements when they are generated and maintained by an electronic record-retention system that meets the following generic criteria: 1) generate and maintain accurate and complete electronic records in a form that does not allow alteration of the record without detection; 2) ensure that the records are not altered throughout the records' required retention period; 3) produce accurate and complete copies of records as required under current regulations; 4) use secure, computer-generated audit trails; 5) ensure that the electronic records are searchable and retrievable; 6) archive electronic records in a form which preserves context, metadata, and audit trail; and 7) make computer systems, controls, and attendant documentation available for agency inspection. When an electronic signature is affixed to an electronic record, the system must also 8) prevent the electronic signature affixed to a document from being detached, copied or otherwise compromised without detection; 9) preserve the basic information associated with electronic signatures (i.e., name, date, time and meaning of affixed signature); and 10) archive an electronic record with affixed signatures in a form which preserves context, metadata, audit trail, and electronic signature.

There are some unique distinctions between the current proposal's approach to electronic reporting and it's approach to electronic record-keeping. For regulated companies submitting environmental compliance reports, the proposed approach relies on an EPA or state controlled system. While for maintaining records electronically, the proposed approach relies on regulated companies to create and operate record-retention systems, select technical approaches, and implement procedures to comply with the rule's performance criteria. Also, with respect to records retained, particularly for third party disclosure purposes, there may be instances where some medium more tangible than electronic records may be preferred. For example, paper may be the medium of choice for record retention of information that must be made available to personnel responding to an emergency, because it may be more likely to remain accessible during emergency events (like power outages, fires, floods, etc.) that could render electronic records inaccessible.

Once the proposed rule is formally published in the *Federal Register Notice*, the regulated community will have an opportunity to formally review it and comment on the approach proposed in the rule. EPA has also announced its plans for further consultation with its stakeholders and to work with the public to promote products that support e-commerce for environmental information.

Conclusion

In recent years, legislative drivers such as the Government Paperwork Elimination Act and the Electronic Signatures in Global and National Commerce Act, have taken the decision to embark on E-Gov away from individual agencies -- mandating the road to E-Gov. Along with new and exciting opportunities to improve our current business processes, E-Gov also brings new challenges, with significant legal and regulatory implications, which agencies like EPA need to address to provide electronic approaches to its reporting community. CROMERRR moves EPA a step closer to E-Gov and shows its commitment to working with its state partners and many stakeholders to effectively address the challenges and fully embrace the opportunities of E-Gov, while maintaining effective stewardship of the environment.

References

1. See *Electronic Reporting at EPA: USEPA Policy on Legal Acceptance of Electronic Submissions,* [Federal Register: September 4, 1996 (Volume 61, Number 172)], pages 46683-46694.
2. USEPA has funded research in this area through grants to such groups as the Environmental Law Institute (ELI) and the National Governors' Association (NGA). See *From Pens to Bytes: Summaries of Court Decisions Related to Electronic Reporting*, June 1999, Environmental Law Institute, ISBN #0-911937-94-3.
3. EPA Fact Sheet on CROMERRR available at http://www.epa.gov/cdx/.
4. The PRA requires federal government entities to obtain approval from OMB to collect any information from the public. OMB may disapprove, approve, or place conditions on the information collection. The purpose is to ensure that collections are not unnecessarily conducted and that the public burden for approved collections is minimized. The 1995 Paperwork Reduction Act took effect on Oct.1, 1995, superseding the PRA of 1980, as amended in 1986. Final Rule issued August 29, 1995 (60 FR 44978).
5. While EPA's earlier efforts focused on larger enterprises submitting electronic reports via value added networks in EDI based standards, more recent efforts have focused on alternative technologies, particularly those attractive to smaller and medium size businesses. For an example of one of EPA's efforts to develop e-reporting options which meets the needs of small and medium-sized businesses, see the report on Internet Discharge Monitoring Report (DMR) pilot with NY Department of Environmental Conservation in 2000 at http://www.epa.gov/cdx.

6. *Access America*, available at Http://www.gits.gov. *Reinventing Environmental Regulation*, President Bill Clinton and Vice President Al Gore, March 16, 1995.
7. U.S. EPA *REI Action Plan: Building Data Systems for the 21st Century*, December, 1997.
8. *The Government Paperwork Elimination Act* (GPEA), Public Law No. 105-277, §§1701-1710 (1998), took effect on October 21, 1998.
9. OMB guidance to assist agencies in implementing GPEA is available at: Http://www.whitehouse.gov/WH/EOP/OMB, Procedures and Guidance; Implementation of the Government Paperwork Elimination Act, 65 FR 25508, May 2, 2000.
10. *The Electronic Signatures in Global and National Commerce Act* (E-Sign), Public Law No. 106-229, enacted on June 30, 2000, codified at 15 U.S.C. §§7001 to 7031.
11. E-Sign defines uniquely governmental transactions as reporting and record-keeping by regulated entities that is principally for governmental purposes. These requirements, while not addressed by E-Sign, are addressed by GPEA. Governmental transactions that are also commercial transactions may be affected by E-Sign. For such transactions, the effective date for E-Sign, for agencies that have undertaken a rulemaking to address electronic reporting/record-keeping, is June 1, 2001. Where E-Sign's applies to an EPA compliance requirement to retain a record, the effective date is June 1, 2001. At the time of this writing, federal agencies were still assessing the impact of E-Sign on regulatory requirements.
12. *EPA's Strategic Plan*, EPA 190-R-00-002, September 2000, page 55.
13. National Governors' Association, *A State Guide for electronic Reporting of Environmental Data,* November 1999, page 63.
14. U.S. Department of Justice, *Legal Considerations In Designing And Implementing Electronic Processes: A Guide For Federal Agencies*, November, 2000, page 7.
15. The current CROMERR proposal does not address electronic reporting on any form of magnetic media (diskette, tape, etc.). The proposed rule does not prohibit such technologies; it simply does not address them.
16. Delegated programs is used throughout this paper to refer to those states authorized to implement the requirements of federal environmental laws. Such state programs are variedly referred to as delegated, authorized, approved, or assumed programs.

17. A concern for a number of states is the proposal's provisions for approving state electronic reporting and record-keeping programs. Many states view the current process for program modifications under existing state primacy regulations as carrying a high administrative cost. Also, of concern to some states is what they perceive as system complexity/cost driven by federal enforcement concerns regarding electronic signatures. For a fuller discussion of state concerns, see NGA *SEES* document.

Chapter 30

Establishing the Groundwork for FIFRA/TSCA Reporting under Electronic Submissions, Signatures, and Record-Keeping

Francisca E. Liem, Mark J. Lehr, and Robert L. Cypher

Laboratory Data Integrity Branch, U.S. Environmental Protection Agency, Mail Stop 2225A, 401 M Street, S. W., Washington, DC 20460

Under the Government Paperwork Elimination Act of 1998, every Federal agency must provide electronic reporting by 2003 as an option for those entities it regulates (1). What this means to the regulated community is that businesses may choose to send electronic information via the Internet instead of mailing or faxing paper copies of reports ranging from Toxics Release Inventory (TRI) forms to Good Laboratory Practice (GLP) study data filed in accordance with FIFRA/TSCA requirements.

EPA is proposing changes to its regulations that would permit the use of electronic document receiving systems to accept electronic records that satisfy certain document submission requirements in Title 40 of the Code of Federal Regulations. EPA believes that electronic filing of information will streamline the submission process for businesses and EPA, as well as state environmental regulators, by eliminating the inconvenience and logistical problems associated with storing mountains of paper records. It is hoped that electronic reporting will increase productivity in the government and private sector alike which will lead to cost reduction. EPA also believes that electronic reporting will improve the quality of records and raw data by eliminating rekeying and allowing for automated quality control. "Today's proposal would set fourth the criteria against which an electronic document receiving system would be evaluated before it could be certified to receive electronic documents in satisfaction of federal environmental reporting or document submission requirements" *(2)*. The new proposal would also establish necessary conditions by which the EPA will identify an electronic record as one that meets all federal record-keeping requirements. It should be stated again that under today's proposal, electronic document submission or electronic record-keeping would be totally voluntary: EPA would not require the submission of electronic document or maintenance of electronic records in lieu of paper documents or records.

History

To begin the discussion concerning EPA's own version of the electronic records and signatures rule...generally known today as **Cro**ss **M**edia **E**lectronic **R**eporting and **R**ecords **R**ule or (CROMERRR), we should first begin with a quick history of the original concept. The Food and Drug Administration (FDA) took the lead in 1990 with the electronic records and signatures rule, originally an initiative, established for the pharmaceutical industry. The new rule paved the way for interested pharmaceutical companies to use electronic signatures rather than hand written ones. The aim of this proposal was to take advantage of technology and avoid the need for paper.

In March 1997 the final rule was published in the Federal Register *(3)* and was effective August 20, 1997. The rule affects both existing and newly developed systems and covers not only electronic signatures but also electronic records.

The major issue from the FDA perspective is that electronic records and electronic signatures must be trustworthy, reliable, and subject to FDA inspection. As with any rule or regulation, the contents outlined in the preamble and the final rule itself represent minimum requirements for implementation. Similar to all regulations and guidelines, the requirements go in the direction of more stringent interpretation.

Requirements For Electronically Maintained Records

Like the FDA, EPA is also concerned about the overall quality of electronic records. EPA is establishing a set of requirements that must be met by the regulated community who wish to maintain electronic records in lieu of paper records. The new regulations will guide regulated entities who use electronic systems to modify, maintain, or transmit electronic records. Regulated facilities would need to employ procedures and controls designed to meet the minimum criteria in today's rule. These are designed to insure that electronic records are trustworthy and reliable, available to agencies as required under current regulations, and will be admissible as evidence in a court of law to the same extent as would be a corresponding paper record.

The EPA's CROMERRR proposal states that for electronic records to be trustworthy, reliable and fully meet agency requirements for record keeping their corresponding electronic record-retention system must establish the following:

1. The ability to generate and maintain accurate and complete copies of records and documents in a form that does not allow alteration of the record without detection.
2. Protection of records without alteration throughout the records retention period.
3. The ability to produce accurate and complete copies of an electronic record and render these copies readily available, both in human readable and electronic form, in normal course of all business processes in a timely manner, as required by predicate regulations, throughout the entire retention period.
4. The ability to ensure that any record bearing an electronic signature contains the name of the signatory, the date and time of signature, and any information that explains the meaning affixed to the signature.
5. The protection of electronic signature so that the signature that has been affixed to a record cannot be detached, copied or otherwise compromised.
6. Use of computer-generated, time stamped audit trails to independently record the date and time of operator entries and actions that create, modify, or delete electronic records. An audit trail is an important element of any acceptable electronic record. It provides an electronic record of key entries and actions to a record throughout the life cycle of the record. Such audit trail documentation needs to be retained for a period at least as long as that required for the subject electronic records. Audit trail documentation also needs to be available for agency review.

7. Records are searchable and retrievable for reference and secondary uses, including audits, legal proceedings, third party disclosures, as required by predicate regulations, throughout the entire retention period.
8. Electronic records must be archived in an electronic form which preserves the context, meta data, and audit trail. Depending on the record retention period required in predicate regulations, regulated entities must insure that complete records, including the related meta data, can be migrated to a new system as needed.
9. Computer systems (including software and hardware) controls, and attendant documentation are readily available for agency inspection.
10. Copies must be maintained for records associated with reports that have been electronically submitted to the agency's Central Data Exchange or to an agency certified electronic report receiving system.

Electronic Records and Electronic Signatures

To ensure that an electronic signature is bound to a responsible individual(s) or business in such a way that no one can repudiate the record, adequate controls must be in place. The EPA's CROMERRR proposal reports that for records with signature requirements which must be maintained but not submitted to the EPA or other authorized agency, electronic records may be used in lieu of paper records when, in addition to the general criteria discussed above, the following conditions are met.

1. The signed electronic records must contain information associated with the signing that clearly indicates the name of the signer, the date and time when the electronic record was signed, and the meaning associated with the signature (such as review, approval, responsibility, authorship, etc.)
2. The electronic signatures must be linked to their respective electronic records to ensure that the signatures cannot be excised, copied or otherwise transferred so as to falsify an electronic record by ordinary means.
3. This information must be subject to the same controls as those for electronic records and must be included as part of any human readable form of the electronic record.

Each of these requirements is necessary to identify the individual during the normal course of business and for unambiguously binding an individual to an electronic record and its content.

Definitions

Understanding some of the basic terms used with the proposed rule is an important step to fully understanding the concepts of electronic records and signatures. We begin first by defining the FDA's new rule for electronic record as "any combination of text, graphics, data, audio, pictorial, or other information presented in digital form that is created, modified, maintained, archived, retrieved, or distributed by a computer system" *(4)*. One must remember that a print out of a computer file is not the actual electronic record itself. Electronic records usually have search, sort, and other capabilities built into it that we do not usually see on a printout. In addition, electronic records have other relevant data, such as which user made a change, the time and date that the change was made, and so on. Summarized in Table I are additional terms some of which will be focused on in more detail later.

Table I. Relevant Terms

	Relevant Terms
Closed System	System access is controlled by information technology people responsible for the content of electronic records
Digital Signature	Electronic, cryptographic, verifiable, signature using a set of rules and parameters that identify the signer...Public Key Infrastructure (PKI).
Electronic Record	Any combination of text, graphics, data, audio, pictorial, or other information representation in digital form that is created, modified, maintained, archived, retrieved, or distributed by a computer
Electronic Signature	A computer data compilation of any symbol or series of symbols executed, adopted, or authorized by an individual that is the legally binding equivalent of the individual's handwritten signature
Open System	An environment in which system access is not controlled by people who are responsible for the content of electronic records that is on the system.

Open and Closed Systems

In determining the nature of electronic systems, one must consider all aspects, including the application, operating system, network, hardware, and all associated services. As indicated in Table I, the main difference between an open and closed system is simply access. "If, for example, you have a chromatography data system that operates within your department, it is a closed system. It is still a closed system if the information technology department runs the server and maintains the network, and it remains a closed system if you out source the information technology support to a third party provider, provided no other company's work interferes with yours. When you start working across the Internet then the chromatography data system becomes an open system, and the FDA rule requires controls" *(5)*.

Again, restating the FDA definition of Electronic Records as any combination of text, graphics, data, audio, pictorial, or other information representation in digital form that is created, modified, maintained, archived, retreated, or distributed by a computer system. From this definition, the laboratory chromatography data system used as the example generates electronic records. Based upon the definition, laboratories will need to consider more than just the raw data files. One must also include the method files, run sequence files, and the integration parameters used for the data analysis. The need for a comprehensive audit trail is a critical component of the FDA regulations and will be discussed in more detail later. The audit trail is also an electronic record that is subject to the same controls.

In addition, an electronic record must consist of two components, a human-readable section and a machine (computer)-readable section. The content of the human-readable section will include information about the creation and any additional processing of the data. A laboratory must have controls, such as, cyclic redundancy checks, to ensure the integrity of any data held within any electronic record.

Audit Trail

One of the key components of the Good Laboratory Practice regulations in ensuring the trustworthiness of data is an effective audit trail, and the same holds true of electronic records and signatures. The FDA regulations require that the audit trail must be computer generated, not a paper record, and that it is an internal part of the application you are using. If you are using a laboratory information management system (LIMS), most of the commercial systems will have an audit trail in which changes are logged behind the scenes and appear only when users must enter the reason for a change. The audit trail has some specific requirements. It must be independent of the operator and cover the lifetime of any electronic record from creation through modification and deletion. When an audited change is made, the audit trail must record:

- Who made the change
- When the change was made including the date and local time in hours and minutes
- The original data without overwriting
- The new entry
- Reason for the change

The audit trail must be retained with, and as long as, the original electronic records. In addition, the audit trail must be in an appropriate form for review or copying by Federal investigators.

Storage Media Issues

We have all been witnesses to the fast-paced evolution of technology, and as a result, it is realistic to expect that electronic records will be transferred from one media format to another during the required period of retention. While EPA has allowed for such transfers in today's proposed rule, it is important to mention that in the spirit of the GLPs, any such transfer must occur in a fashion that ensures that the entire electronic record is preserved without modification. As noted earlier, the electronic record includes not only the electronic document itself, but also the required information regarding time and date of receipt, etc. "Any method of migrating electronic records from one electronic storage medium to another that fails to meet this criterion will not produce records that meet Federal environmental record-retention requirements. For example, a CD-ROM version of a record originally stored on electromagnetic tape would not satisfy Federal record-keeping requirements unless the method for transferring the record from one medium to the other employed error-checking software to ensure that the data were completely and faithfully transcribed. The Agency is currently seeking comment on whether this criterion is sufficient to ensure that the integrity and authenticity of the electronic record is maintained throughout its required record retention period" *(2)*.

Central Data Exchange (CDX)

One of the primary differences between the FDA's CFR Part 11 and EPA's CROMERRR is the creation of the Central Data Exchange or CDX. The Agency's Office of Environmental Information (OEI) is currently developing the specifications for a Central Data Exchange (CDX) that will serve as the Agency's gateway for electronic documents received by the EPA. CROMERRR reports that with respect to the electronic document submission addressed by today's proposal, CDX functions will include:

- **Access management** - allowing or denying an entity access to CDX
- **Data interchange** - accepting and returning data via various file transfer mechanisms
- **Signature/certification management** - providing devices and mandatory scenarios for individuals to sign and certify what they submit
- **Submitter and authentication** - assuring that signatures are valid and data are uncorrupted
- **Transaction logging** - providing date, time, and source information for data received to establish chain of custody
- **Acknowledgment and provision of copy of record** - providing the submitter with confirmations of data received
- **Archiving** - placing files received and transmission logs into secure, long-term storage
- **Error-checking** - flagging obvious errors in documents and document transactions, including duplicate documents and unauthorized submissions
- **Translation and forwarding** - converting submitted documents into formats that will load to EPA databases and forwarding them to the appropriate systems
- **Outreach** - providing education and other customer services to CDX

The idea is to provide "one way and one place" for the regulated community to exchange electronic documents with EPA. CDX may also provide the platform for state-EPA data exchanges. However, as with the provisions of the proposed rule, the features and functions of CDX described will generally be inapplicable to these state-EPA exchanges.

The use of the CDX will require little more than access to a computer with a browser and Internet connection. However, for organizations that have invested heavily in the computerized management of their environmental data, CDX is also designed to support substantial automation of the data transfer processes. In addition, the EPA hopes that CDX's centralization of data exchanges will eventually provide the platform for greater integration or consolidation of environmental reporting. To support the various functions of CDX, a number of components will have to be incorporated which include:

- Digital signatures based on public key infrastructure (PKI) - PKI is a way of reliably establishing and maintaining the identity of the individual producing digital signatures.
- A process for registering users and managing their access to CDX - EPA would require entities to register with CDX prior to electronic data submission. Initially this will be by invitation from EPA.
- A characteristic system architecture - EPA has been guided by three goals in designing the CDX:

  - *Flexibility in exchanging data* - support several data exchange mechanisms.
  - *Uniformity in signing/certifying submissions* - uniform way for individuals to sign and certify their electronic documents.
  - *Adequate security for all aspects of CDX operation* - Authorized users of CDX, including EPA, retain control over the CDX operations for which they are responsible

- Electronic Data Interchange (EDI) standards - transmission of electronic data in a standard syntax, of unambiguous information between computers of organizations.
- A characteristic environment in which electronic reporting transactions will be conducted - CDX will allow the submitter to transmit data either through automated file transfer, or via on screen "smart forms" provided as a part of the downloaded "desktop".

EPA believes these building blocks, taken together, do satisfy the criteria that today's proposal specifies for electronic documents receiving systems.

Enforcement

Much has changed over the past twenty two years within the world of Good Laboratory Practices, and no one area has seen more change than that of information technology. This varying landscape has forced government to modify existing interpretations of regulations, and to develop new regulations such as those proposed in the recent consolidation effort. Among the proposed changes to the new regulations will be sections focusing on laboratory technology. These new subsections will address issues concerning the integrity of data stored and manipulated by computers, data processors, and automated laboratory procedures. In addition, the proposed regulations will focus on current "state-of-the-art" issues including electronic records and signatures. But until the new regulations become final, GLP investigators will need to apply existing regulations to the electronic records and signatures rule. Table II provides some typical citations that an investigator might apply to a facility using an electronic data system in place of a manual recording system.

By publishing the final rule EPA hopes that CROMERRR will accomplish the following: "allow the agency to comply with Paper work Elimination Act, provide a uniform, technology-neutral framework for electronic reporting and record keeping, allow EPA programs to offer electronic reporting and record keeping as they become ready, provide states with a uniform set of criteria for approval of their electronic reporting and record keeping provisions, and ensure that electronic reporting and record keeping under EPA and State-EPA programs do not compromise the enforceability of environmental programs" *(2)*. EPA's final rule will be consistent with FDA's electronic records rule (21 CFR Part 11).

Conclusion

As EPA moves toward the eventual finalization of its own rule, currently scheduled for release in 2001, the pressure will mount for EPA investigators and the regulated community to learn of the benefits of the electronic records and signatures rule, as well as the potential consequence resulting from lack of compliance. Like the FDA, the EPA will look to see whether a company has identified a plan for bringing their systems and procedures into compliance, whether they have performed an assessment or gap analysis, whether they have the necessary controls in place, and whether they are systematically bringing their systems into compliance. EPA investigators expect to handle any potential rule violation as they handle any other deviation with the current GLP regulations. Each deviation will be looked at individually to determine its nature and extent, its impact on product quality or data integrity, and finally the company's compliance history. EPA is certain that with time and patience, the electronic records and signatures rule will become an integral part of doing business that will save time, money and resources as we continue into the new millennium.

Table II. Possible GLP Citations Applying to Electronic Data Systems

\multicolumn{2}{c}{*Possible FIFRA/TSCA GLP Citations That Apply to Electronic Data Systems*}	
Good Laboratory Practice Citation	**Interpretation**
40 CFR 160/792.29(a) - Personnel	Personnel not trained to operate electronic data collection system
40 CFR 160/792.33 - Study Director	Problems with GLP compliance for a study always comes back to the study director
40 CFR 160/792.35(b)(5) - QAU	Unauthorized deviations to the SOPs or protocol
40 CFR 160/792.61 - Equipment Design	Data system is not of appropriate design or adequate capacity to function according to protocol and method requirements
40 CFR 160/792.63(b)(c) - Maintenance & Calibration of Equipment	Lacking SOPs on maintenance, calibration, inspection, or testing of data systems or part of systems
40 CFR 160/792.81(a) - Standard Operating Procedure (SOP) - General	Possible deviations from laboratory SOPs on data systems, lack of SOPs for the use and generation of electronic data capture.
40 CFR 160/792.81(b)(10) - Standard Operating Procedure (SOP) - Data Handling	Lacking required SOPs on data handling, storage, and retrieval
40 CFR 160/792.120(a)(13) - Protocol (Maintaining Records)	Protocol must identify records to be maintained
40 CFR 160/792.130(e) - Conduct of Study	No audit trail for changes to data Audit trail established, however; Cthe actual change was not made Coriginal entry was obscured Cthe reason for change was not stated Cindividual making the change was not identified No individual identified for direct data input

References

1. Hogue, Cheryl. "Paperless Regulators: EPA Prepares For Electronic Reporting", *Chemical and Engineering News,* American Chemical Society, September 4, 2000.
2. CROMERRR Proposal: "Establishment of Electronic Reporting; Electronic Records" (United States Environmental Protection Agency, July 31, 2000).
3. Code of Federal Regulations, Food and Drugs, Title 21, Part 11 "Electronic Records, Electronic Signatures" (U.S. Government printing Office, Washington D.C. 1997).
4. Motise, P.; Part 11: "Electronic Records, Electronic Signatures - Answers to Frequently Asked Questions," Presented at the FDA Electronic Records Workshop, Philadelphia, 12 January 1999.
5. Dowall, R.D.; "Just e-sign on the Bottom Line?" *LCGC*, February 2000, Volume 18, Number 2.

Chapter 31

Considerations for Data Collection Systems

Rodney M. Bennett

Cerexagri, Inc., 900 First Avenue, King of Prussia, PA 19406

As our economic and social world becomes more and more globalized, the information that is required just to have a chance at survival is increasing. We all are being forced to find the most efficient and cost effective means to gather, correlate, store and retrieve many types of information that are generated on a day-to-day basis. This information must be in a highly usable form and easily cross-referenced. The data must be in both complete and partially retrievable formats. In our regulated society, information must be in a verifiable form, such that the generation and correlation source(s) are easily tracked and identified. In addition, sufficient media must be generated such that the information remains immutable throughout an extended lifetime. The electronic media allow for this type of collection, cross-referencing and correlation in an environment that remains verifiable and immutable over time (to the extent of our current and changing knowledge base.) The information presented in the ACS short course evolved into an exchange of ideas and questions from which this collection of considerations was compiled. This chapter gives an outline

Based upon comments from an open discussion Short Course presented by Rodney M. Bennett in conjunction with the ACS Symposium entitled: "Capturing and Reporting Data in the Electronic Age" of possible items which may help in determining a proper course of action in the daily struggle to collect, process, store and retrieve one of our most important assets: DATA.

Background

The collection of information and the ability to use that information has been the need of both animals and plants over the centuries. The collection and storage of information comes in various forms. The most basic in life is the DNA code. In nature, some changes in storage can be a good thing, as in altered DNA code, which may allow diversity. The data that we collect under a regulated environment must be able to be maintained in an unchanged form. In the field or analytical laboratory, the data must be able to be collected in an efficient manner. The type of collection media used will be dependent upon the type of data being collected as well as the conditions under which it is being collected. In order to determine the type of collection system that may be needed for a particular type of data, a series of stepwise decisions should be made. These will include the basic steps in the interrogatory process: WHO?, WHAT?, WHEN?, WHERE?, WHY? and finally HOW?

The WHO?

In most organizations, it is best to establish a group or committee to address the data collection process. The group should be well focused and charged with specific duties to perform. The group should include persons from the various disciplines or functions that will be responsible in the generation of the data, review of the data, storage of the data and oversight of the data collection process. These individuals will include: 1) the on-line producers of the data (the chemist/technologist in the field and/or analytical, toxicological, medical, etc., laboratory); 2) the data collection and storage groups (information service groups, system managers); and 3) the data integrity groups (quality assurance, quality control, system managers). The group does not have to be large. The group should have sufficient technical knowledge or access to that knowledge to be able to determine the types of data that will be collected, the manner in which the data will be used, and the way the data and the electronic collection system will be maintained.

A protocol should be established for the process. The protocol will include: the purpose and scope of the project; the topics to be addressed; the systems to which the protocol will apply; and the specific requirements for the systems being evaluated. A protocol will require each member of the group to follow a set outline and keep the group focused on specific goals. As with any protocol, amendments may be required as the project progresses. The protocol should not be too restrictive. New and creative ideas should be encouraged.

A specific timeline should be established for the group. The timeline should be such that ALL critical events are met. For an initial group, the first meeting may be to determine IF an electronic data system is needed. In most laboratories, this answer will be YES, due to the type of data that is being collected. In the past, data were recorded by hand in notebooks or on pre-printed forms. This type of data collection still has its place. Over the years, we have found that this information will need to be correlated and summarized into some type of report format. At this point, an additional "pen and paper", typewritten or computer generated format is produced. The data then must be copied and distributed, and the original data must be stored. Storage of some types of data is limited and costly due to the storage media. When the information is used for some later evaluation, "pen and paper", typewritten or computer printed reports must be retrieved. Data and evaluations require the transfer of information manually to another medium. The use of an electronic medium reduces the amount of transfer of data and the possible transcription errors associated with the transfer. Once a system is established, another WHO question will arise. WHO will have the authorization to do WHAT? This should be tailored to the individual company. Some suggestions are presented later in this chapter. Now, WHAT do we need to get started?

The WHAT?

The WHO group will be faced with a multitude of questions. The first will be several "What" questions. The first "What" question may be: What are the costs and benefits with putting in this system? This question will be answered in various and different ways depending upon the type of company. Does your company really need to have such a validated, compliant system? The cost issue will be determined primarily by whether your answer is: *"Yes"* or *"No"*. If you are not in a regulated environment, your initial answer may be "No", but you will want to consider, in this case, the advantages and cost savings that may be seen by having the data in a consistent, secure and auditable format. If you are in a regulated environment, your initial answer is "Yes", if you want to remain in your current business! Your benefits are many. Among these benefits are: 1) continued ability to produce and sell your product; 2) avoid regulatory penalties or fines; 3) avoid delay in registration of a product; and 4) consistency and ease of storage and retrieval of data. There will be many other questions that will need to be answered. What type of data do we need to collect? What instruments will be involved in the collection, evaluation and storage of the data? What is the best form(s) for these data? What access will be needed for the data in the future? What groups of people will be using the data? What are the storage requirements? What tracking and verification systems will be needed? What systems do we already have in place? What are the costs versus benefits for installing a compliant system? What systems or media exist externally that may be used? Which components if any do we implement? Do we put these into practice now or wait for additional instruction from the regulators and defer the costs? In most instances, the basic system should be installed as soon as possible. Many regulations are already in place for record keeping. These regulations cover data submitted for the Organisation for Economic Co-operation and Development (OECD) *(1)* as well as all data submitted under Good Laboratory Practice Standards for the U.S. Environmental Protection Agency (EPA) *(2)* and the US Food, Drug, and Cosmetic Act (FDA). The new data collection and storage systems will allow the users to address these regulations in a more cost-effective manner. If you wait until a new regulation is in place, such as the OECD GLP Consensus Document No. 19 *(3)* or the USEPA "Good Automated Laboratory Practices" *(4)* and have not had some contingency planning ahead of time, you may not have sufficient time to incorporate a proper system. The costs of reliable systems also may increase. What do we do FIRST? and WHEN?

The WHEN?

The evaluation group must establish a logical course of action. The protocol will help to focus the group and help to establish the timeline. The group will need to decide when the group will meet and how often? When does the system need to be in operation? For the data system, when are the data collected? When does a data transfer occur? When are the data interpreted? When are the data stored? When are the data used? When do we START this process of evaluation? Then, WHERE is this leading?

The WHERE?

The group will need to establish a number of "Where" questions. Where are the data being generated? Where do we want to have access allowed? Where do we want to store the data? Where do we want to house the collection and storage equipment? Where are the internal and external resources that will be required to properly evaluate and implement the system? Where do we Start? And some may ask: "WHY are we going through this process?"

The WHY?

Some of the reasons for conducting the evaluation process have already been discussed. If you or any member of the team cannot answer "Why" they are doing the evaluation, seek help from upper management. You may have answered your own question or you may need to get additional members on the evaluation group. In some instances, the "WHY" question may indicate that the specific items that have been established in the protocol have missed the mark. A "Why" question that arises from a requirement standpoint should always be evaluated. Why have we always collected data in this manner? Why is our equipment set up in this manner? Why do we want to have access or permissions of a particular type established for specific individuals? Why don't we have a look at "HOW" we can best approach this system.

The HOW?

The group will want to know "How" to start the evaluation process. The development of the protocol will start the process. How many goals are needed? How many systems do we need to evaluate? How are we going to manage the process once it is established? How will we know when we have completed the process? This last question is easy. This will be a continuing process that will need to be evaluated from time to time on a regular basis for the lifetime of the system. Once the system is established, standard operating procedures (SOPs) will be required, and these SOPs will need to be reviewed and updated on a regular basis. Now that we have all this philosophy established and so many questions unanswered, "HOW" and "WHERE" do we go from here? And WHO is going to give us a start?

The Nuts and Bolts

First we start with the protocol. The protocol will establish the type of system(s) that will be used. The protocol may be written for the entire system or parts of the system. Many software packages such as the Agilent ChemStation ® (5) and the Waters Millennium ® (6) systems have both stand-alone instrument controller versions as well as network versions. There are several Operating Systems under which the various programs run. There are a variety of Operating Systems which include Windows 95 ® (7), Windows 98 ® (7) and Windows NT ® (7) Operating Systems. The Windows NT ® Operating System has some advantages in having a higher degree of security than some of the other systems. The security of the system is an important factor to determine for any operating system that is chosen. A careful evaluation of the available systems should be conducted by the investigation team. The recommendation will be brought to the group or to management. The specific reasons for the particular system of choice and its advantages, limitations and ability for future upgrades and expansion should be discussed. Once a system is chosen, a protocol should be established on "How the System will be Used and Maintained". The protocol may include the installation and upgrading parameters as well as general operating criteria as defined below.

The protocol will define 1) the specific system(s) and its components; 2) installation and validation parameters for the system(s) and its components; 3) "How" and "By Whom" changes to the system(s) and its components are made; 4) documentation and audit trail requirements for the system(s) and its components; and 5) archival considerations for software, data and other associated media. The system and associated components for a single unit system may include 1) the instrument, detector(s) and auto-sampler(s); 2) any external switching device(s) or collection device(s); 3) the associated computer(s), printer(s), monitor(s), integrator(s), and any other data detection or collection devices; and 4) storage device(s). The system may be associated with a network. A network occurs if any two or more computers are connected together and are able to collect and exchange data. Networks may be in the form of a local area network, which may include one of more instruments or computers (generally of the same type) in the same general area. They may be fully integrated networks, which connect multiple instrument(s) of various types over multiple locations. The locations can range from within a specific building or on a specific campus site to global connections for multiple labs worldwide. If the system is under a local area network, a single server computer would be added to this list along with any other controller units and associated network software programs and peripherals. For a fully integrated network, additional computers, collection, transfer and storage devices are added. These may include 1) the network server; 2) remote access and controller devices; and 3) conversion boxes for analog/digital conversions. In many multiple vendor systems, the software access codes are not widely distributed such that an analog signal generated from a specific instrument must be converted to a digital signal for data acquisition, data processing and data storage. A variety of HUB and Switch type devices also are generally associated with signal transfer to a centralized network system.

There should be a description of the system as to the location of its components and their associated connections. This will include but not be limited to 1) the manufacturer; 2) model names/numbers; 3) serial numbers (hardware, software and peripherals); 4) operating systems and their versions; and 5) application software and their versions. All licensing and software/hardware documentation and validation should be archived as soon as possible. A working copy should be kept in the laboratory to be used on a day-to-day basis.

A back-up system and back-up procedures should be established. Many back-up systems can either be performed in real time or on a routine maintenance schedule. The back-up procedures may be performed manually or established as an automated system. The back-up media should have a long life expectancy (such as storage on a CD type system). The archival procedure should be specified in a SOP. A contingency for a system failure ("crash") and recovery procedures should be specified.

The privileges and responsibilities for each user and management should be clearly defined. Many systems have a built-in list of privileges that may be assigned to various groups and may be specified by management. For a single-user system, individual maintenance by trained associates may be acceptable. For small networked systems, it is highly desirable to have a System Administrator to oversee maintenance of the system, particularly the software and hardware connection devices. For a medium or large network, a System Administrator is essential.

In most of the data systems, the privileges are defined based upon the level of access that the specific individual requires. An "Audit Trail" log should be an essential part of any system used in a regulated environment. The "Audit Trail" should log the date and time of any access event to the system including:

system assignments; method creation; data acquisition; data review; data calculations; data storage; and any other activity which is performed on the system. This should be a secure log that cannot be altered or purged without high level authority. A SOP should be established on how any deletions and/or any changes to the Audit Trail will be made, by whom they may be made, and how they will be tracked. Training is essential at all permission levels including a "View Only" level. A "View Only" level may be appropriate for a Quality Assurance function. The privileges should be consistent with the specific functions for that individual user. An analyst (technician) may be given only the privileges that will allow him or her to load samples, use a pre-set method for an instrument, and start the analysis run. This individual cannot change any of the pre-set values for the method and may or may not be allowed to actually enter the sample listing. The analyst often can only look at his or her own method and data and does not have access to other methods, data or instruments. A chemist level typically will have privileges to use multiple methods and may have the privilege to modify a method. The chemist may or may not have the privilege to add an instrument to the system or change the configuration of the individual components of the system. In most cases, the chemist will have access to the data evaluation portion of the software and will be allowed to process the data. The degree to which the data processing may be adjusted by the user will vary. The chemist will typically have the ability to view more than one method and may be allowed to view other colleagues' methods and data. The next supervisor level (project leader/group leader/manager) will typically be given the ability to view data and systems from one or more groups.

Typically, the supervisor will have the privilege to adjust the method(s), change the calculation and processing methods as well as the reporting methods. Many additional levels of privileges may be added. The highest level will be the System Administrator. The System Administrator will have the overall responsibility to ensure that the data system is in working order, to create privileges and profiles for individual users and user groups, to create and add systems and system components, and to back-up and verify system integrity. The System Administrator can change passwords that allow access to the systems and can remove privileges as required by management. Only the System Administrator should have the privilege to DELETE information. This would include instrument methods; collection methods; processing methods; reporting methods; raw data; processed data; all tracking logs; and system components. The protocol for the system should specify the tracking requirements and back-up procedures and precautions that will be taken prior to any deletions from the system and how these will be tracked, recorded and archived. Often it is good to have a Back-Up System Administrator who can function as a replacement if the System Administrator is out for an extended period of time or in the event of an emergency or failure of the system. Many systems will keep a permanent record of all users even if all privileges are removed (as would occur when an individual leaves the group or company). In the case of the System Administrator, it is often good practice to require at least two people to be able to remove the System Administrator privileges. Often the systems will allow a single System Administrator to remove anyone, but the two-person requirement may be added as an SOP or as a part of the system protocol.

The system must be validated in some manner. System validation has been discussed in greater detail in other chapters in the book. The main point for validation is this: When evaluating a system and writing the protocols and SOPs, make sure that the type of validation that you will require is understood by you and your vendor. The vendor typically conducts the software system validation and provides some certification. The system integrity then is verified in the laboratory during installation. Check with the vendor to determine the upgrade policies to the system and when and who would conduct the verification of any upgrades or modifications to the system. Another consideration is the calculation parameters and validation/verification of the calculation and report routines. Often the vendor will have some standard materials that may be run under specific conditions to determine some general parameters. There are often pre-prepared data sets within the system, which also can help in the verification process. Validation of the system data collection and processing software is typically the responsibility of the vendor. The end user then evaluates specific data under various conditions and under varied calculation and reporting parameters to establish a verified method for analysis. Some of the parameters that are typically established for any analytical method will be the heating conditions, flow rates (for gases or liquid mobile phases), attenuation, baseline adjustments; and data fit of known concentrations of standards (establish a standard curve) for linear, quadratic or multiple order fits. Also, methods typically will have an associated Limit of Quantitation (LOQ) and a Limit of Detection (LOD), which may be listed in the method notes. These are all method specific and should be generated on an individual method basis and stored under a specific method name.

A list of required entries should be established for each analysis run. The required entries may be listed as a part of any subsection(s) of the method. Typically, data systems will establish: an instrument method; a calculation method; a sample set listing; a sequence listing (which shows the order of injection for the samples); a processing method; and a reporting method. Many systems also include printing and storage options. The list of required items often will include but will not be limited to 1) a record number for the study or analysis being conducted; 2) each individual sample identification (and in some systems a sample sequence of analysis); 3) the system(s) being used for the analysis; 4) the instrument method; 5) the method of calculation: 6) the name of the individual conducting the analysis; 7) the date that the data are acquired; 8) the date that the data are processed and by whom they are being processed; and 9) the method of reporting. There may be more than one processing date and associated processor of the data; therefore, all are recorded and logged.

All exporting of the data should be tracked and a log (audit trail) kept to indicate when, where, why and by whom any data are transferred. All system, software and instrument changes should be kept under strict "change control" parameters that are specified in appropriate protocols and SOPs.

All users of the system should be fully trained to their access privilege level. An in-house System Administrator or authorized supervisor may conduct initial training for limited users. Higher level training, especially for the System Administrator, should be conducted by the manufacturer/vendor of the system.

Once the system is in place, there will always be changes to the system as the needs of the company grow and as various regulations are put into place. Hopefully, these general tips will help in the preparation of an initial outline for your system needs and provide a guide for future expansion.

References

1. Organisation for Economic Co-operation and Development, 21 January 1998. OECD Series on "Principles of Good Laboratory Practice and Compliance Monitoring".
2. Organisation for Economic Co-operation and Development, 1995. OECD GLP Consensus Document No. 10. "The Application of the Principles of the GLP to Computerised Systems".

3. U.S. Code of Federal Regulations (CFR), 40, part 160, 1989 (revised July 01, 1991). "Good Laboratory Practice Standards."
4. U.S. Environmental Protection Agency, 1995. "Good Automated Laboratory Practices." EPA Directive 2185.
5. Agilent Technologies (formerly Hewlett-Packard Corp.), P.O. Box 954, Santa Clarita, CA 91380-9972 USA.
6. Waters Corporation, 34 Maple Street, Milford, MA 01757 USA.
7. Microsoft Corporation (Windows Software), P.O. Box 1096, Buffalo, NY 14240-1096 USA.

Appendix

Establishment of Electronic Reporting: Electronic Records

Proposed Rule

U.S. Environmental Protection Agency

EPA is proposing to allow electronic reporting to EPA by permitting the use of electronic document receiving systems to receive electronic documents in satisfaction of certain document submission requirements in EPA's regulations. The proposal also sets forth the conditions under which EPA will allow an electronic record to satisfy federal environmental recordkeeping requirements in EPA's regulations. In addition, under today's proposal, States and tribes will be able to seek EPA approval to accept electronic documents or allow the maintenance of electronic records to satisfy reporting and recordkeeping requirements under authorized or delegated environmental programs that they administer. The proposal includes criteria against which a State's or tribe's electronic document receiving system will be evaluated before EPA can approve changes to the authorized program to allow electronic reporting. Similarly, the proposal includes criteria against which EPA will evaluate a State's or tribe's provisions for electronic recordkeeping. Under today's proposal, electronic document submission or electronic recordkeeping will be totally voluntary; EPA will not require the submission of electronic documents or maintenance of electronic records in lieu of paper documents or records. EPA will only begin to accept direct submission of an electronic document once EPA has provided public notice that its electronic document receiving system is prepared to receive the document in electronic form. Similarly, EPA will only begin to allow electronic records to satisfy a specific EPA recordkeeping requirement once EPA has provided public notice stating that electronic records will satisfy the identified requirement.

Supplementary Information

Affected Entities

This rule will potentially affect State and local governments which have been authorized or which seek authorization to administer a federal environmental program under Title 40 of the Code of Federal Regulations. The rule will also potentially affect private parties subject to any requirements in Title 40 of the Code of Federal Regulations that a document be created, submitted, or retained. Affected entities include the examples listed in Table I.

This Table is not intended to be exhaustive, but rather provides a guide for readers regarding entities likely to be affected by this action. This Table lists the types of entities that EPA is now aware can potentially be affected by this action. Other types of entities not listed in the Table can also be affected. Note that while this proposal will affect entities involved with hazardous waste management, it does not apply to the Hazardous Waste Manifest, which EPA is addressing in a separate electronic reporting rule.

Preamble

Table of Contents

Information in the preamble is organized as follows:

I. Overview

    A. Why does the Agency want to allow electronic reporting and record-keeping?

Table I. Examples of Affected Entities

Category	Examples of Affected Entities
Local government	Publicly owned rreatment works, owners and operators of treatment works treating domestic sewage, local and regional air boards, local and regional waste management authorities, municipal and other drinking water authorities.
Private	Industry owners and operators, waste transporters, privately owned treatment works or other treatment works treating domestic sewage, privately owned water works, small businesses of various kinds, sponsors such as laboratories that submit or initiate/support studies, and testing facilities that both initiate and conducts studies.
State government	States or Tribes that manage any federal environmental programs authorized/approved by EPA under Title 40 of the Code of Federal Regulations.
Federal government	Federally owned treatment works and industrial dischargers, federal facilities subject to hazardous waste regulation.

B. What will the proposed regulations do?

II. Background

    A. What is EPA's current electronic reporting policy?
    B. How will today's proposal change EPA's current electronic reporting policy?
    C. Why is EPA proposing these changes in electronic reporting policy?
    D. What is EPA's approach to electronic record-keeping?
    E. What information is EPA seeking about electronic reporting and record-keeping proposals?
    F. How were stakeholders consulted in developing today's proposal?

III. Scope of Today's Proposal

    A. Who may submit electronic documents and maintain electronic records?
    B. How does today's proposal relate to the new E-SIGN legislation?
    C. Which documents can be filed electronically?
    D. Which records can be maintained electronically?
    E. How will today's proposal implement electronic reporting and record-keeping?

IV. The Requirements in Today's Proposal

    A. What are the proposed requirements for electronic reporting to EPA?
    B. What requirements must electronically maintained records satisfy?

        1. General approach.
        2. EPA's proposed criteria for electronic record-retention systems.
        3. Electronic records associated with electronic signatures.
        4. The relation of these requirements to Food and Drug Administration (FDA) criteria under 21 CFR part 11.
        5. Storage media issues.
        6. Additional options.

    C. What is the process that EPA will use to certify State systems as functionally equivalent to the CDX?
    D. What criteria are EPA proposing that State electronic report receiving systems must satisfy?

        1. General system-security requirements.

2. Electronic signature method.
3. Submitter registration process.
4. Electronic signature/certification scenario.
5. Transaction record.
6. System archives.

E. What are the costs and benefits associated with today's proposal?

V. The Central Data Exchange (CDX)

A. What is EPA's concept of the CDX?
B. What are the CDX building blocks?

1. Public key infrastructure (PKI)-based digital signatures.
2. The CDX registration process.
3. The CDX architecture.
4. Electronic data interchange (EDI) standards.
5. The transaction environment.

VI. Regulatory Requirements

A. Executive Order 12866
B. Executive Order 13132
C. Paperwork Reduction Act
D. Regulatory Flexibility Act
E. Unfunded Mandates Reform Act
F. National Technology Transfer and Advancement Act
G. Executive Order 13045
H. Executive Order 13175
I. Executive Order 13211 (Energy Effects)

Topic Discussion

I. Overview

A. Why Does the Agency Want To Allow Electronic Reporting and Record-Keeping?

More than ten years ago, EPA published a notice entitled: "Electronic Reporting at EPA: Policy on Electronic Reporting," (FRL-3815-4) announcing the goal of making electronic reporting available under EPA regulatory programs. We gave as reasons for this goal our expectation that enabling the submission and storage of electronic documents in lieu of paper documents can:

- Reduce the cost for both sender and recipient,
- Improve data quality by automating quality control functions and eliminating rekeying, and
- Greatly improve the speed and ease with which the data can be accessed by all who needed to use it.

Electronic reporting and record-keeping have a strong mandate in federal policy and law. As stated in the March, 1996, Reinventing Environmental Information Report, electronic reporting supports the President's overall regulatory re-invention goals of reducing the burden of compliance and streamlining regulatory reporting. In addition, the Government Paperwork Elimination Act (GPEA) of 1998, Public Law 105-277, requires that agencies be prepared to allow electronic reporting and recordkeeping under their regulatory programs by October 21, 2003. Given the enormous strides in data transfer and management technologies since 1990--particularly in connection with the Internet--replacing paper with electronic data transfer now promises increased productivity across almost all facets of business and government.

B. What Will the Proposed Regulations Do?

The proposed rule will remove existing regulatory obstacles to electronic reporting and record-keeping across a broad spectrum of EPA programs, and establish requirements to assure that electronic documents and electronic records are--for all purposes--as valid and authentic as their paper counterparts. These proposed requirements will apply to regulated entities that choose to submit electronic documents and/or keep electronic records, and under today's proposal, the choice of using electronic rather than paper for future reports and records will remain purely voluntary. Today's proposal will not amend compliance requirements under existing regulations and statutes and will not affect whether a document must be created, submitted, or retained under the existing provisions of Title 40 of the Code of Federal Regulations. Similarly, today's proposal will not affect the period of required record-retention, whether the stored electronic document must be signed, who is entitled to receive copies of the record, the number of copies that must be maintained, or any other requirements imposed by the underlying EPA, State, tribal or local program regulations. Public access to environmental compliance information will not be adversely affected by today's proposal. Electronic reporting and record-keeping provisions in this proposal will provide for continued public access to electronic documents equivalent to that provided for paper records under existing law.

For purposes of this proposal, EPA is using the term "electronic" reportingin a sense that excludes submission of a report via magnetic media, for example via diskette, compact disk, or tape; we are also excluding transmission via hard copy facsimile or "fax". Likewise, our use of the term "electronic document" throughout this Notice refers exclusively to documents that are transmitted via a telecommunications network, excluding hard copy facsimile. However, this proposal's exclusion of magnetic media submissions in no way indicates EPA's rejection of this technology as a valid approach to paperless reporting; we believe that in many cases magnetic media submission fulfills the goals of the Government Paperwork Elimination Act (GPEA). Many EPA programs have successfully used magnetic media submissions to implement their regulatory reporting, including Hazardous Waste, Toxic Release Inventory, and Pesticide Registration. EPA expects these magnetic media approaches to paperless reporting to continue, and nothing in today's proposal should be understood to proscribe them.

For regulated entities that choose to submit electronic documents directly to EPA, today's proposal will require that these documents be submitted to a centralized Agency-wide electronic document receiving system, called the `Central Data Exchange' (CDX), or to alternative systems designated by the Administrator. Regulated entities that wish to submit electronic documents directly to EPA will satisfy the requirements in today's proposal by successfully submitting their reports to the CDX. While we do not intend to codify any of the details of how CDX operates or how it is constructed, EPA does solicit comments on the characteristics of the CDX and the submission scenarios described in this preamble. In addition, the CDX design specifications will be included as a part of this rulemaking docket. For regulated entities that choose to keep records electronically, today's proposal requires the adoption of best practices for electronic records management. Importantly, today's proposal will not authorize the conversion of existing paper documents to an electronic format for record-retention purposes because no mechanism currently exists that can be relied upon in all cases to preserve the forensic data in an existing paper document when it is converted to an electronic form. However, today's proposal does not prohibit such conversions at the Administrator's discretion on a case-by-case basis.

Many facilities do not submit documents directly to EPA, but rather to States, tribes or local governments that are approved, authorized or delegated to administer a federal environmental program on EPA's behalf or to administer a state environmental program in lieu of the federal regulatory program in that State. We will refer to these as "authorized State and tribal programs." This proposal will allow for EPA approval of changes to authorized State and tribal programs to provide for electronic reporting, and EPA approval will be based largely on an assessment of the State's or tribe's "electronic document receiving system" that will be used to implement the electronic reporting provisions. For this purpose, today's proposal includes detailed criteria that EPA will use to determine that an electronic document receiving system is acceptable. These criteria address such issues as system security, the approach to electronic signature and certification, chain-of-custody and archiving, including provisions that address how a State, tribe or local government manages electronic records that are directly associated with its electronic document receiving system, as well as certain data transfers between this system and regulated entities. Beyond this, today's proposal does not address State, tribal or local government electronic recordkeeping or data transfers carried out to administer their authorized programs. Today's proposal does not address any data transfers between EPA and States or tribes as a part of administrative arrangements to share data. Finally, it is worth noting that EPA can approve changes to authorized State or tribal programs that involve the use of CDX to receive data submissions from their regulated communities. CDX has been designed with the goal of fully satisfying the criteria that this proposal specifies for assessing State or tribal electronic document receiving systems; similarly, EPA will ensure that other systems the Administrator designates to receive electronic submissions will satisfy the criteria as well. In view of this, EPA is exploring opportunities to leverage CDX resources for use by States, tribes and local environmental agencies.

Similarly, many facilities maintain records to satisfy the requirements of authorized State and tribal programs. This proposal will also allow for EPA approval of changes to authorized State and tribal programs to provide for electronic record-keeping. EPA approval in this case will be based on a determination that the State's or tribe's program will require best practices for electronic records management, corresponding to EPA's provisions for electronic records maintained to satisfy EPA recordkeeping requirements.

For both document submission and record-keeping, the point of the proposed requirements is primarily to ensure that the authenticity and integrity of these documents and records are preserved as they are created, submitted, and/or maintained electronically, so that they continue to provide strong evidence of what was intended by the individuals who created and/or signed and certified them. Among other things, today's proposal is intended to ensure that the federal laws regarding the falsification of information submitted to the government still apply to any and all electronic transactions, and that fraudulent electronic submissions or record-keeping can be prosecuted to the fullest extent of the law. In establishing clear requirements for electronic reporting systems and electronic records, this proposed rule will help to minimize fraud by assuring that the responsible individuals can be readily identified.

While today's proposal will remove regulatory obstacles to electronic reporting and record-keeping, EPA will make electronic submission available as an option for specific reports or other documents only as the systems become available to receive them. Similarly, EPA will make electronic recordkeeping available as an option for specific record-keeping requirements only as programs become ready to adopt this change. In the case of electronic reporting, EPA plans to move aggressively toward implementation of CDX for high volume environmental reports submitted directly to EPA. EPA will publish announcements in the Federal Register as CDX and other systems become available for particular environmental reports and as programs become ready to make electronic recordkeeping an option. These points are discussed in more detail in Section III.C and D of this Preamble. To implement electronic reporting and recordkeeping under authorized State and tribal programs, EPA also plans to work with interested States and tribes to approve the necessary program changes as quickly and expeditiously as possible.

II. Background

A. What Is EPA's Current Electronic Reporting Policy?

On September 4, 1996, EPA published a document entitled "Notice of Agency's General Policy for Accepting Filing of Environmental Reports via Electronic Data Interchange (EDI)" (61 FR 46684) (hereinafter referred to as "the 1996 Policy"), where "EDI" generally refers to the transmission, in a standard syntax, of unambiguous information between computers of organizations that may be completely external to each other (61 FR at 46685). This notice announced our basic policy for accepting electronically submitted environmental reports, and its scope was intended to include any regulatory, compliance, or informational (voluntary) reporting to EPA via EDI.

In the context of EDI, the "syntax" of the computer-to-computer transmissions may be thought of as the structure or format of the transmitted data files. And, ``format" here refers to such things as the ordering and labeling of the individual elements of data, the symbol used to separate elements, the way that related elements are grouped together, and so on. For example, for a file consisting of people's names, a simple format specification might be that (i) the elements occur in order: first-name, middle-name, last-name; (ii) the elements are labeled, respectively, "F", "M", and "L"; (iii) each group of first, middle and last names is separated by a semi-colon; and (iv) there is a comma between any two elements in a group.

For purposes of the 1996 policy, the standard transmission formats used by EPA were to be based on the EDI standards developed and maintained by the American National Standards Institute (ANSI) Accredited Standards Committee (ASC) X12. By linking our approach to the ANSI X12 standards, we hoped to take advantage of the robust ANSI-based EDI infrastructure already in place for commercial transactions, including a wide array of commercial off-the-shelf (COTS) software packages and communications network services, and a growing industry community of EDI experts available both to EPA and to the regulated community. At the time EPA was writing this policy, ANSI-based EDI was arguably the dominant mode of electronic commerce across almost all business sectors, from aerospace to wood products, at least in the United States. EDI was also widely used in the Federal Government, most notably at the Department of Defense, but also, increasingly, at other agencies, including the Social Security Administration, the General Services Administration, the Department of Transportation, the Health Care and Finance Administration, and the Department of Housing and Urban Development, and the Department of Health and Human Services.

However, as the 1996 policy made clear, no specific EPA reporting requirement can be satisfied via EDI until the Agency develops the corresponding program-specific implementation guidance (61 FR 46686). This guidance generally needs to do at least three things. First, it needs to address such procedural matters as the interactions with the communications network (for EDI purposes, usually stipulated as a controlled-access, "value-added network" or "VAN"), schedule for submissions and acknowledgments, transaction records to be maintained, and so on. Second, it needs to stipulate the specific ANSI X12 standard transmission formats--referred to as "transaction sets"--to be used for the specified reports. This stipulation is essential, since ANSI provides hundreds of different transaction sets, each corresponding to a distinct type of commrcial document, e.g., invoices, purchase orders, shipping notices, product specifications, reports of test results, and so on. Third, the guidance also needs to say how the stipulated transaction sets are to be interpreted. X12 transaction sets are generally designed to be somewhat generic--they typically leave a number of their components as "optional", and use data-element specifications that are open to multiple interpretations. (For a more detailed explanation of EDI and these implementation guidance documents, see section V.B.4 of this preamble.)

Given a public notice that the applicable implementation guidance is ready, the September, 1996, policy allows facilities to submit required reports electronically using EDI once they enter into a Terms and Conditions Agreement (TCA) with the Agency (61 FR 46685). Where the report in question requires a responsible individual at a facility to certify to the truthfulness of the submitted data, the TCA must provide for the use of a

Personal Identification Number (PIN) as a form of electronic signature. Under the policy, the individual entering into the TCA is required to use a PIN assigned by EPA for this purpose (61 FR 46685). Finally, under the TCA, the facility is required to adhere to security and audit requirements as described in the notice (61 FR 46687).

Finally, the 1996 policy also explained that the various programs may require additional security procedures on a program-by-program basis (61 FR 46684). Such procedures may be covered in the program-specific implementation guidance, or can be provided through rule-making.

B. How Would Today's Proposal Change EPA's Current Electronic Reporting Policy?

For practical purposes, the most important changes that today's proposal makes to current policy is in our technical approach to electronic reporting. Generally, we propose to greatly broaden the options available for electronic submission of data. For example, while we will continue to support data transfer via standards-based EDI (as explained in section V.B.4 of this preamble) we will also provide options involving user-friendly "smart" electronic forms to be filled out on-line, on the Internet, or downloaded for completion off-line at he user's personal computer. In addition, we propose to support data transfers through the Internet, via email, or via on-line interactions with Web sites, in a variety of common application-based formats, such as those output by spreadsheet packages. In terms of electronic signature technology, while we may continue to allow PIN-based approaches, our plan is to emphasize digital signatures based on "public key infrastructure" (PKI) certificates, given the increasing support for--and acceptance of--PKI for commercial purposes. (For an explanation of PKI, see Section V.B.1 of this preamble.) And, we plan to consider and allow for other signature technologies as they become viable for our applications.

This proposal also represents some important changes in EPA's regulatory strategy as well. To begin with, we are proposing to abandon any attempt to use regulations or formal policies to place technology-specific or procedural requirements on regulated entities submitting electronic documents. In place of the technology-specific/procedural provisions, our regulation will require that electronic submissions be made to designated EPA systems, or to State, tribal or local government systems that are determined to satisfy a certain set of function-based criteria. Thus, as a rulemaking, today's proposal will govern electronic reporting by placing requirements on the systems that receive the electronic documents--rather than on the regulated entities submitting them-- and by specifying these requirement in terms of technology-neutral functionality.

This new regulatory strategy does not mean that we are proposing to abandon any control over how electronic documents are submitted. We are proposing instead to require the use of the "Central Data Exchange" (CDX) system or other EPA designated systems for submissions to EPA. While the rule may be technology-neutral, CDX itself will incorporate a suite of very specific technologies, including digital signatures based on "public key infrastructure" (PKI) certificates, described in detail below. In addition, while the rule itself will not require more than the use of CDX for electronic submissions to EPA, using CDX will--as a practical matter--impose a very well-determined set of requirements on the reporting process for those who choose electronic submission instead of paper when reporting directly to EPA. Section V of this preamble will describe these requirements in some detail.

These changes in strategy are significant. They represent a decision that the mechanics of electronically submitting data should not be reflected in specific regulatory provisions. In addition, these changes give EPA the flexibility to adapt our electronic reporting systems to evolving technologies without having to amend our regulations with each technological innovation. That is, CDX or other designated systems can be changed as appropriate, so long as they continue to satisfy the function-based criteria that the rule establishes. In general, we believe that this strategy will enable EPA, the States and tribes to offer regulated companies a very user-friendly approach to electronic reporting that can be tailored to the level of automation they wish to achieve, and can incorporate improved technologies as they become available without the delay associated with rulemaking.

C. Why Is EPA Proposing These Changes in Electronic Reporting Policy?

EPA is proposing these changes for three reasons. First, and most important, the technology environment has changed substantially since the September, 1996, policy was written. Web-based electronic commerce and Public Key Infrastructure (PKI) provide two obvious examples. While both were available and in use for some purposes in 1996, they had not yet achieved the level of acceptance and use that they enjoy today. We could not have anticipated in 1996 that this evolution would occur as rapidly as it has. Clearly, these developments require that we extend our approach to electronic reporting beyond EDI and PINs. In addition, they teach us that it is generally unwise to base regulatory requirements on the existing information technology environment or on assumptions about the speed and direction of technological evolution.   Second, we believe that technology-specific provisions would, of necessity, be very complex and unwieldy. The resulting regulation would likely place unacceptable burdens on regulated entities trying to understand and comply with it, and might also be difficult for EPA to administer and enforce.

Third, and finally, an electronic reporting architecture that makes a centralized EPA, State or tribal system the platform for such functions as electronic signature/certification is now quite viable--and quite consistent with the standard practices of Web-based electronic commerce. In many ways, regulated entities' electronic transactions with the "Central Data Exchange" (CDX) will be similar to doing business with an on-line travel agency, book store, or brokerage, and with a similar client-server architecture. Given the state of technology five years ago, we could not have considered this approach in the September, 1996, policy.

D. What Is EPA's Approach to Electronic Record-Keeping?

Today's proposal sets forth the criteria under which the Agency considers electronic records to be trustworthy, reliable, and generally equivalent to paper records in satisfying regulatory requirements. The intended effect of this proposed rule is to permit use of electronic technologies in a manner that is consistent with EPA's overall mission and that preserves the integrity of the Agency's enforcement activities.

E. What Information Is EPA Seeking About Electronic Reporting and Record-Keeping Proposals?

In proposing to allow regulated entities to submit electronic documents and maintain electronic records, EPA has, at least, the following three goals:

- To reduce the cost and burden of data transfer and maintenance for all parties to the data exchanges;
- To improve the data--and the various business processes associated with its use--in ways that may not be reflected directly in cost-reductions, e.g. through improvements in data quality, and the speed and convenience with which data may be transferred and used; and
- To maintain or improve the level of corporate and individual responsibility and accountability for electronic reports and records that currently exists in the paper environment.

EPA is seeking comment and information on how well today's proposed regulatory provisions and the associated Central Data Exchange infrastructure will serve to fulfill these three goals. Concerning the first--addressing cost and burden--EPA is particularly interested in and seeks comment on whether today's proposal will make electronic reporting and record-keeping a practical and attractive option for smaller regulated entities, especially small businesses. Concerning the second--addressing the data and the associated business process--we are especially interested in comments on how our proposed approach to electronic reporting and record-keeping will affect third parties, for example State and local agencies that may collect and/or use the data in implementing EPA programs as well as members of the public who have an interest in the data as concerned citizens.

Concerning our third goal, it is essential that we continue to ensure sufficient personal and corporate responsibility and accountability in the submission of electronic reports and the maintenance of electronic records; otherwise we place at risk the continuing viability of self-monitoring and self-reporting that provides the framework for compliance under most of our environmental programs. Therefore, EPA is especially interested in any concerns or issues that commenters may wish to raise about the effect that moving from paper to the electronic medium may have on this compliance structure--as well as assessments of the approaches EPA is proposing to address these concerns.

F. How Were Stakeholders Consulted in Developing Today's Proposal?

Today's proposal reflects more than eight years of interaction with stakeholders--including State and local governments, industry groups, the legal community, environmental non-government organizations, ANSI ASC X12 sub-committees, and other federal agencies. Many of our most significant interactions involved electronic reporting pilot projects conducted with State agency partners, including the States of Pennsylvania, New York, Arizona, and several others. In addition, over a two-year period beginning in May, 1997, EPA worked together with approximately 35 States on the State Electronic Commerce/Electronic Data Interchange Steering Committee (SEES) convened by the National Governors' Association (NGA) Center for Best Practices (CBP). The product of the SEES effort was a document entitled, ``A State Guide for Electronic Reporting of Environmental Data,'' available in the docket for this rulemaking, along with reports on some of the more recent state/EPA electronic reporting pilots. Information on SEES is also available at: www.nga.org/CBP/Activities/EnviroReporting.asp. Today's proposal has benefitted greatly from the SEES discussions, and EPA believes that the proposal is generally consistent with the SEES ``State Guide''.

Beginning in June, 1999, EPA also sponsored a series of conferences and meetings with the explicit purpose of seeking stakeholder advice on today's rulemaking. These included:

- The Symposium on Legal Implications of Environmental Electronic Reporting, June 23-25, 1999, convened by the Environmental Law Institute;
- Two NGA-convened State meetings, held in Cleveland, April 11-12, 2000, and in Phoenix, June 1-2, 2000; and
- Two public meetings, held in Chicago, June 6, 2000, and in Washington, D.C., July 11, 2000.

Reports of these conferences and meetings are also available in the rulemaking docket.

III. Scope of Today's Proposal

    A. Who May Submit Electronic Documents and Maintain Electronic Records?

Any regulated company or other entity that submits documents addressed by today's proposal (see section III.B., below) directly to EPA can submit them electronically as soon as EPA announces that the Central Data Exchange or a designated alternative system is ready to receive these reports. Any regulated company or other entity that maintains records addressed by today's proposal (see section III.C., below) under EPA regulations can store them in an electronic form subject to the proposed criteria for electronic record-keeping as soon as EPA announces that the specified records may be kept electronically. As noted in section I.B of this preamble, the rule will not authorize the conversion of existing paper records to an electronic format. Regulated companies or other entities that submit documents or maintain records under authorized State or tribal programs may submit or maintain them electronically as soon as EPA approves the changes to the authorized programs that are necessary to implement the State's or tribe's provisions for electronic reporting or recordkeeping.

Under today's proposal, the entities that can use electronic reporting and record-keeping will not be required to do so; they can still use the medium of paper for document submissions and records if they choose. Nonetheless, nothing in this proposal will prohibit State, tribal or local authorities from requiring electronic reporting or record-keeping under applicable State, tribal and local law.

    B. How Does Today's Proposal Relate to the New E-SIGN Legislation?

The environmental reports and records that are the subject of this rule are generally not subject to the recently enacted "Electronic Signatures in Global and National Commerce Act of 2000" ("E-SIGN" or "the Act"), Public Law 106-229, because most of these governmentally-mandated documents are not amongst the "transactions" to which E-SIGN applies. However, the EPA has authority to permit electronic reporting under the statutes it administers and under the Government Paperwork Elimination Act (GPEA) of 1998, Public Law 105-277, http://ec.fed.gove/gpedoc.htm. E-SIGN, establishes the legal equivalence between: (1) Contracts written on paper and contracts in electronic form; (2) pen-and-ink signatures and electronic signatures; and (3) other legally-required written documents (termed "records" in the statute) and the same information in electronic form. As a general rule, if parties to a transaction in interstate commerce choose to use electronic signatures and records, E-SIGN grants legal recognition to those methods. E-SIGN provides that no contract, signature, or record relating to such a transaction shall be denied legal effect solely because it is in electronic form, nor may such a document be denied legal effect solely because an electronic signature or record was used in its formation. GPEA also provides such language for government filings covered by this rule and provides similar legal validity for associated electronic signatures. When E-SIGN takes effect on October 1, 2000, statutes or agency rules containing paper-based requirements that might otherwise deny effect to electronic signatures and records in consumer, commercial or business transactions between two or more parties will be superseded. E-SIGN does, however, permit federal and State agencies to set technology-neutral standards and formats for the submission and retention of electronic documents.

E-SIGN applies broadly to commercial, consumer, and business transactions in or affecting interstate or foreign commerce, including transactions regulated by both federal and State government. However, the conferees who drafted this legislation specifically excluded "governmental transactions" from the definition of transactions that are subject to E-SIGN; accordingly, E-SIGN does not cover transactions that are uniquely governmental, such as the transmission of a compliance report to a federal or State agency. Nonetheless, E-SIGN does cover documents that are created in a commercial, consumer, or business transaction, even if those documents are also submitted to a governmental agency or retained by the regulated community for governmental purposes. For example, an insurance contract that is commemorated in an electronic document will be covered by the provisions of E-SIGN, even if EPA or an authorized State requires that the policy-holder maintain proof of insurance as part of a federal or State environmental program. In order to ensure that these documents will meet governmental needs, the Act permits the government to set technology-neutral standards and formats for such records. In order that governmental agencies have time to promulgate these standards and formats, E-SIGN has a delayed effective date for its record-retention provisions of March 1, 2001. If a federal or State regulatory agency has proposed a standard or format for document retention by March 1, 2001, the Act will take effect with respect to those records on June 1, 2001.

C. Which Documents Could Be Filed Electronically?

With the exception of the Hazardous Waste Manifest (which EPA is addressing in a separate electronic reporting rule), today's proposal addresses document submissions required by or permitted under any EPA or authorized State, tribal or local program governed by EPA's regulations in Title 40 of the Code of Federal Regulations (CFR). Nonetheless, EPA will need time to develop the hardware and software components required for each individual type of document. Similarly, EPA will need time to evaluate State, tribal, and local electronic document receiving systems to ensure that they meet the criteria articulated in today's proposal. Accordingly, once this rule takes effect, documents subject to this rule submitted directly to EPA can only be submitted electronically after EPA announces in the Federal Register that the Central Data Exchange (CDX) or an alternative system is ready to receive them. Documents subject to this rule submitted under an authorized State or tribal program can only be submitted electronically once EPA has approved the necessary changes to the authorized program.

Both in developing the CDX, and in approving changes to authorized State and tribal programs related to electronic reporting, EPA plans to give priority to receipt of the relatively high volume environmental compliance reports that do not involve the submission of confidential business information (CBI). EPA believes that receipt of electronically transmitted CBI requires considerably stronger security measures than the initial version of CDX may be able to support, including provisions for encryption. While EPA does plan to enhance CDX to accommodate CBI, we will first want to gain experience implementing CDX in the non-CBI arena and also take the time to explore CBI security issues with companies that submit confidential data. EPA seeks comments and advice on priorities for electronic reporting implementation. EPA also seeks comments on this proposal's global approach, and whether specific exclusions should be added to the rule.

D. Which Records Can Be Maintained Electronically and Which Can Not?

Today's proposal addresses records that EPA or authorized State, tribal or local programs require regulated entities to maintain under any of the environmental programs governed by Title 40 of the CFR or related State, tribal and local laws and regulations. Nonetheless, individual EPA programs may need additional time to consider more specific provisions for administering the maintenance of electronic records under their regulations. Similarly, EPA will need time to evaluate State, tribal, and local programs' provisions for administering electronic records maintenance to ensure that such records will meet the criteria articulated in today's proposal.

Accordingly, once this rule takes effect, any records subject to this rule submitted directly to EPA can only be maintained electronically after EPA announces in the Federal Register that EPA is ready to allow electronic records maintenance to satisfy the specified record-keeping requirements. Records subject to this rule maintained under an authorized State or tribal program can only be maintained electronically once EPA has approved the necessary changes to the authorized program. For electronic records specified in such Federal Register announcements or authorized program changes, they can be maintained in lieu of paper records so long as they meet the requirements in this proposal, unless paper records are specifically required in regulations promulgated on or after promulgation of this final rule. However, today's proposal will not apply to paper records that are already in existence--whether these are maintained under EPA programs or under authorized State, tribal or local programs--and will not provide that any of these paper records can be converted to an electronic format. In addition, today's proposal does not address contracts, grants, or financial management regulations contained in Title 48 of the CFR. EPA is addressing such procurement-related activities separately. Accordingly, today's proposal does not apply to records maintained under these Title 48 regulations, whether this record-keeping was administered by EPA or by a State, tribal or local program under EPA authorization.

E. How Would Today's Proposal Implement Electronic Reporting and Record-Keeping?

EPA proposes our overall policy and requirements for electronic reporting and record-keeping as a new 40 CFR part 3, which consists of four (4) Subparts. Subpart A provides that any reporting requirement in Title 40 can be satisfied with an electronic submission to EPA that meets certain conditions (specified in Subpart B) once EPA publishes a notice that electronic document submission is available for this requirement. Similarly, Subpart A provides that any record-keeping requirement in Title 40 can be satisfied with electronic records that meet certain conditions (specified in Subpart C) once EPA publishes a notice that electronic record-keeping is available for this requirement. Subpart A also provides that electronic reporting and record-keeping can be made available under EPA-authorized State, tribal or local environmental programs as soon as EPA approves the necessary changes to these authorized programs (in accordance with Subpart D). In addition, subpart A makes clear: (1) That electronic document submission or record-keeping, while permissible under the terms of this part, will not be required; and (2) that this regulation will confer no right or privilege to submit data electronically and will not obligate EPA or State, tribal or local agencies to accept electronic data except as provided under this regulation.

Subpart B sets forth the general requirements for acceptable electronic documents submitted to EPA. It provides that electronic documents must be submitted either to EPA's Central Data Exchange (CDX) or other EPA designated systems. It also includes general requirements for electronic signatures. Subpart C sets forth requirements that regulated entities must satisfy if they wish to maintain their electronic records in satisfaction of EPA record-keeping requirements. Finally, subpart D sets forth the process and criteria for EPA approval of changes to authorized State, tribal and local environmental programs to allow electronic document submissions or record-keeping to satisfy requirements under these programs. With respect to electronic document submissions, subpart D includes detailed criteria for acceptable State, tribal or local agency electronic document receiving systems against which EPA will assess authorized program implementations of electronic reporting. Table II describes the applicability of each of these proposed new subparts.

Given the proposed provisions of Subpart A, a regulated entity wishing to determine whether electronic reporting or record-keeping was available under some specific regulation will have to verify that EPA has published a Federal Register notice announcing their availability and will have to locate any additional provisions or instructions governing the electronic option for the particular reporting or record-keeping requirements. EPA seeks comments on whether the new Part 3 should include specific cross-references to such announcements and instructions to the extent that these are codified elsewhere in Title 40. The cross references could be organized by CFR subparts of Title 40, and could provide a simple listing of program-specific regulations for which EPA has implemented electronic reporting or record-keeping under the provisions of today's proposal. EPA invites suggestions on the most helpful cross-referencing scheme.

IV. The Requirements in Today's Proposal

    A. What Are the Proposed Requirements for Electronic Reporting to EPA?

Today's proposal specifies just two requirements for electronic reporting to EPA. First, electronic documents must be submitted to an appropriate EPA

Table II. Applicability of Proposed Subparts

Subpart	Applicability
A. General Provisions	Companies and other entities regulated under Title 40 of the Code of Federal Regulations, and State, tribal and local agencies with electronic document receiving systems used to receive documents under their authorized programs.
B. Electronic Reporting to EPA	Companies and other entities regulated under Title 40 of the Code of Federal Regulations.
C. Electronic Record-keeping Under EPA Programs	Companies and other entities regulated under Title 40 of the Code of Federal Regulations.
D. Approval of Electronic Reporting & Record-keeping Under State Programs	State, tribal and local agencies with electronic document receiving systems or electronic record-keeping programs for which EPA approval is required.

electronic document receiving system; generally this will be EPA's Central Data Exchange (CDX), although EPA can also designate additional systems for the receipt of electronic documents. Second, where an electronic document must bear a signature under existing regulations or guidance, it must be signed (by the person authorized to sign under the current applicable provision) with an electronic signature that can be validated using the appropriate EPA electronic document receiving system. The proposal stipulates that the electronic signature will make the person who signs the document responsible, or bound, or obligated to the same extent as he or she would be signing the corresponding paper document by hand. Only electronic submissions that meet these two requirements will be recognized as satisfying a federal environmental reporting requirement, although failure to satisfy these requirements will not preclude EPA from bringing an enforcement action based on the submission.

It should be noted that the second requirement, concerning signatures, will apply only where the document would have to bear a signature were it to be submitted on paper, either because this is stipulated in regulations or guidance, or because a signature is required to complete the paper form. Today's proposal is not intended to require additional signatures on documents when they are migrated from paper to electronic submission. The EPA electronic document receiving system will indicate to the submitter whether a signature is required to complete submission of an electronic document--although the presence or absence of this indication will not affect whether or not a signature is required for a document to have legal effect.

Beyond these two requirements, the proposed rule does not specify any required hardware or software. Accordingly, the proposed rule text does not include any detail about CDX per se or about what will be required of regulated entities who wish to use it. Nonetheless, in publishing today's proposal, one of EPA's goals is to share our plans for the CDX and to invite comments on the technical approaches that it represents. Therefore, section V, below, explains the details of CDX as it is currently planned--including CDX technical approaches to satisfying our proposed functional criteria, and what use of CDX to submit electronic documents will require of the users. We are also including the draft CDX design specifications in the docket for today's proposed rule. In reviewing these materials, however, the reader should bear in mind that the details of CDX that they specify have not been finalized, and may be affected by the comments received on today's proposal. In the preamble to the notice of final rulemaking for today's proposal, EPA will describe the details of CDX as it will actually be implemented, and will highlight any significant changes from the design as described in this proposal.

Of course, even after the current CDX design is finalized and implemented, the system may change--to take advantage of opportunities offered by evolving technologies, as well as to correct any deficiencies that operational experience reveals. Our proposed regulatory strategy--avoiding the codification of technology-specific/procedural provisions--is meant to accommodate such changes without requiring that we amend our regulations. Nonetheless, EPA recognizes that such changes can be disruptive to regulated entities that participate in electronic reporting; therefore, we are adding provisions that commit EPA to provide adequate public notice where a contemplated change may have this impact. In general, we foresee four kinds of cases:

- Major changes that can be disruptive to regulated entities; these will likely affect the kinds of hardware or software required to submit electronic reports--examples may include required changes to the file formats CDX will accept, or to the required electronic signature technology, but will not generally include optional upgrades to software, the provision of additional formatting (or other technical) options, or changes to CDX that simply reflect changes to the regulatory reporting requirements that the system is supporting;
- Minor changes that will likely not be disruptive; these will affect the user interface but without affecting the hardware or software required to submit electronic reports--examples may include changes to screen layouts, or sequencing of user prompts;
- Transparent changes that will affect CDX operation without any apparent change in interaction with submitters--an example may be a change to the CDX archiving process; and
- Emergency changes necessary to protect the security or operational integrity of CDX--an example may be an upgrade to the system firewall protection.

Our approach will then be to provide public notice and seek comment on major changes at least a year in advance of contemplated implementation. For minor changes we will provide public notice at least 60 days in advance of implementation. For transparent changes and emergency changes we will make decisions on whether and when to provide public notice on a case-by-case basis. EPA seeks comment on this approach, including the kinds of cases we distinguish and the proposed time-frames for notice. We are especially interested in views on the appropriateness of the time-frame for notice of major changes--and specifically on whether a shorter time-frame, e.g. 9 months or 6 monnths, would provide adequate notice while giving EPA greater flexibility to make timely responses to changes in the technological environment. We also seek comment on the more general question of whether it is in the best interests of EPA and our regulated entities to codify these public notice provisions at all, or whether they may place at risk our ability to be sufficiently responsive to the changing needs of our user community. We are also interested in the question of whether the different kinds of cases are or can be defined with sufficient precision to form the basis for workable regulatory provisions, and we welcome any suggestions for alternative regulatory language.

B What Requirements Must Electronically Maintained Records Satisfy?

1. General Approach.

In today's proposed rule, EPA is proposing a set of criteria that will have to be met by regulated entities that maintain electronic records in lieu of paper records, to satisfy record-keeping requirements under EPA regulations in Title 40 of the CFR. The proposed criteria address the minimal functional capabilities that an electronic record-retention system must possess in order for an electronic record or document to meet a federal environmental record-keeping requirement. Regulated entities that use electronic systems to create, modify, maintain, or transmit electronic records will need to employ procedures and controls designed to meet the minimum criteria in today's rule. These criteria are designed to insure that electronic records are trustworthy and reliable, available to EPA and other agencies and their authorized representatives in accordance with applicable federal law, and admissible as evidence in a court of law to the same extent as a corresponding paper record.

2. EPA's Proposed Criteria for Electronic Record-Retention Systems.

In general, EPA believes that for electronic records to be trustworthy and reliable, their corresponding electronic record-retention system must: (1) Generate and maintain accurate and complete copies of records and documents in a form that does not allow alteration of the record without detection; (2) ensure that records are not altered throughout the records' retention period; (3) produce accurate and complete copies of an electronic record and render these copies readily available, in both human readable and electronic form as required by predicate regulations, throughout the entire retention period; (4) ensure that any record bearing an electronic signature contains the name of the signatory, the date and time of signature, and any information that explains the meaning affixed to the signature; (5) protect electronic signatures so that any signature that has been affixed to a record cannot be detached, copied, or otherwise compromised; (6) use secure, computer-generated, time-stamped audit trails to automatically record the date and time of operator entries and actions that create, modify, or delete electronic records; (An audit trail is an important element of any acceptable electronic record, for it provides an electronic record of key entries and actions to a record throughout its life cycle. Such audit trail documentation needs to be retained for a period at least as long as that required for the subject electronic records. Audit trail documentation also needs to be available for agency review.) (7) ensure that records are searchable and retrievable for reference and secondary uses, including inspections, audits, legal proceedings, third party disclosures, as required by predicate regulations, throughout the entire retention period; (8) archive electronic records in an electronic form that preserves the context, metadata, and audit trail; (Depending on the record retention period required in predicate regulations, regulated entities must insure that the complete records, including the related metadata, can be maintained in secure and accessible form on the preexisting system or migrated to a new system, as needed, throughout the required retention period.) and (9) make computer systems (including hardware and software), controls, and attendant documentation readily available for agency inspection. EPA believes that where these 9 criteria are met, records required to be maintained under EPA regulations, can be kept electronically, including where they involve or incorporate signatures.

   3. Electronic Records with Electronic Signatures.

Where electronic records involve or incorporate electronic signatures meeting the requirements under Subpart C of this proposal, EPA will consider the electronic signatures to be equivalent to hand-written signatures. EPA believes the criteria described in paragraph B.2. above address the conditions for cases of electronic records involving signatures, such as: first, a signed electronic record must contain information associated with the signing that clearly indicates the name of the signer, the date and time when the electronic record was signed, and, the meaning associated with the signature (such as review, approval, responsibility, authorship, etc.); second, electronic signatures must be linked to their respective electronic records to ensure that the signatures cannot be excised, copied or otherwise transferred so as to falsify an electronic record by ordinary means; third, this information will be subject to the same controls as those for electronic records and must be included as part of any human readable form of the electronic record (such as electronic display or printout). EPA seeks comment on whether these criteria are appropriate and whether--taken together with the general criteria--they are sufficient to ensure that signatures associated with records fulfill their purpose. EPA also seeks comment on whether these criteria are appropriate for the maintenance of electronic records containing digital signatures. (For an explanation of digital signatures, and their role in CDX, see Section V.B.1 of this preamble.) The special issues involved in maintaining digitally signed records are discussed in Section IV.D.6 of this preamble--in connection with archiving requirements for electronic document receiving systems--and EPA is interested in views on whether these issues need to be more explicitly addressed by the criteria for electronic record-retention systems discussed here, especially the criterion provided in Sec. 3.100(5), which addresses the maintenance of the electronic signature as a part of the electronic record. EPA seeks comment on whether this provision should be expanded to accommodate some of possible procedures for archiving digital signatures referred to at the end of Section IV.D.6.

4. The Relation of These Requirements to Food and Drug Administration (FDA) Criteria.

The criteria set forth in today's proposed rule--both the general and those specific to records with associated signatures--are intended to be consistent with criteria set forth for electronic document systems in other relevant regulations, such as FDA's criteria in 21 CFR part 11. EPA seeks comment on whether today's proposed requirements achieve this consistency, and whether this consistency is an appropriate goal for this rulemaking.

5. Storage Media Issues.

Given the fast-paced evolution of technology, it is realistic to expect that electronic records will be transferred from one media format to another during the required period of record retention. While EPA allows for such transfers in today's propose rule, any such transfer must occur in a fashion that ensures that the entire electronic record is preserved without modification. As noted earlier, the electronic record will include not only the electronic document itself, but also the required information regarding time of receipt, date of receipt, etc. Any method of migrating electronic records from one electronic storage medium to another that fails to meet this criterion will not produce records that meet federal environmental record-retention requirements. For example, a CD-ROM version of a record originally stored on electromagnetic tape will not satisfy federal record-keeping requirements unless the method for transferring the record from one medium to the other employs error-checking software to ensure that the data is completely and faithfully transcribed. EPA seeks comment on whether this criterion is sufficient to ensure that the integrity and authenticity of the electronic record is maintained throughout its required record retention period.

6. Additional Options.

In addition to the criteria discussed above, EPA is currently evaluating the need for additional controls for electronic records under this rule. Over the course of the next five (5) months, EPA plans to conduct additional analysis, and based on the results of this analysis and the public comments received on the electronic record provisions contained in today's proposal, EPA may determine that additional provisions are required for electronic records. If such a determination is made, prior to proposal of the final rule, EPA will publish a supplemental notice detailing any additional electronic record provisions to be included in the final rule. We realize that the electronic records criteria in today's rule are not as detailed as that contained in FDA's 21 CFR part 11 and seeks comments on whether our proposed criteria are sufficient to ensure the authenticity, integrity, and non-repudiation of electronic records maintained by regulated facilities in fulfillment of their compliance obligations. EPA is considering whether or not to include additional provisions found in the FDA regulations in our final rule. Such provisions could include the following: (1) Establishment and implementation of written policies that limit system access to authorized individuals, as well as the use of authority checks to ensure that only authorized individuals can use the system, electronically sign a document, access the operation or computer system input or output device, alter a record, or perform the operation at hand; (2) establishment and implementation of written policies that hold individuals accountable and responsible for actions initiated under their electronic signatures, in order to deter record and signature falsification; (3) use of device (e.g., terminal) checks to determine the validity of the source of data input or operational instruction; (4) use of additional measures such as document encryption and use of appropriate digital signature standards to ensure, record authenticity, integrity, and non-repudiation; (5) routine and documented validation of systems to ensure accuracy, reliability, consistent intended performance, and the ability to discern invalid or altered records; (6) establishment and implementation of written policies governing education and training of personal and certification that persons who develop, maintain, or use electronic record signature systems have the education, training, and experience to perform their assigned tasks. EPA is also seeking comment on the general feasibility of converting existing paper documents-- including litigation-sensitive records--to electronic documents, as well as comments on the strengths and weakness of existing technologies available for this purpose.

C. What Is the Process That EPA Will Use To Approve Changes To Authorized State and Tribal Programs Related to Electronic Reporting and Record-Keeping?

EPA expects that States, tribes and local agencies that administer EPA-authorized environmental programs will wish to implement electronic reporting and recordkeeping at least as quickly and extensively as EPA. Therefore, in overseeing these programs, EPA wishes to balance multiple objectives of minimizing administrative burden on States, providing State flexibility for varying State approaches, and ensuring that State systems are robust enough to meet the demands of a strong enforcement capability. EPA considered several options for meeting these needs, including program-by-program approval processes--in each case under applicable EPA program-specific regulations--State self-certifications, and a centralized approval process. This proposal provides for State flexibility by specifying performance criteria rather than requiring specific technologies, and balances other objectives though use of a hybrid process for approving changes to authorized State and tribal programs.

Under this process, EPA will provide a single set of substantive performance criteria, listed in today's proposal, that will apply to any authorized program where EPA determines that electronic reporting and record-keeping will involve substantive changes to the program that will require EPA approval. Today's proposal contains language that would make compliance with these Part 3 criteria an element of all authorized State, tribal, or local programs that wish to accept electronic reports or allow electronic recordkeeping, although the language does not change the procedural requirements for modifications to any of these program. This means, for example, that a State planning to institute electronic reporting for an authorized program will have to meet the normal EPA approval requirements for that program--whether the approval sought is for a single program or for an electronic document receiving system that would support multiple authorized, delegated, or approved environmental programs. In the case where multiple programs will be affected, the State will still need to seek modification of each such program under existing program approval or revision procedures; however, EPA expects that it will evaluate such multiple applications in a single internal review. Moreover, EPA solicits comment on whether another approach should be taken to State and tribal program modification or revision for electronic reporting or record-keeping.

Alternatively, State, tribal or local agencies may wish to rely on third-party systems to receive reports on their behalf, where these systems are operated or owned by commercial or not-for-profit organizations. Today's proposal will allow this on the condition that the electronic document receiving system employed by the State, tribal or local agency satisfy the substantive performance criteria that we specify, and authorization approvals are obtained where necessary.

D. What Criteria Are EPA Proposing That State Electronic Report Receiving Systems Must Satisfy?

In today's proposed rule, EPA is providing a set of criteria that will have to be met by any system that is used to receive electronic documents submitted to satisfy electronic document submission requirements under any EPA-authorized State, tribal, or local environmental program. The proposed criteria address the functional capabilities that EPA believes a State's, tribe's or local government's ``electronic document receiving system'' must have if it is to ensure the authenticity and non-repudiation of these electronic documents. EPA has developed these criteria to ensure that any electronic document has the same legal dependability as its paper counterparts. EPA does not intend to imply that information or documents derived from electronic reporting or record-keeping systems that do not meet all of EPA's criteria, or from transactions that were not in compliance with all applicable requirements and agreements, could not be introduced as evidence at trial, would not constitute admissions, or would not constitute records required by, or used for compliance with, applicable statutes (e.g., Clean Water Act section 309(c)(4), Resource Conservation and Recovery Act section 3008(d)(3)). EPA's criteria are intended to result in systems and records that will provide the best evidence for use by plaintiffs and prosecutors in enforcement actions, and to facilitate the success of such enforcement actions. These criteria are designed to ensure any electronic document used as evidence in the course of prosecuting an environmental crime or civil violation will have the same or better evidentiary value as its paper equivalent. For example, the criteria are designed to ensure that in prosecuting the crime of deliberate falsification of compliance data, the identity of the person who signed a falsified document can be established beyond a reasonable doubt. One of the criteria, entitled "Validity of Data," and proposed in section 3.2000(b), addresses this standard directly. In general, a system that is used to receive electronic documents must be capable of reliably generating proof for use in private litigation, enforcement proceedings, and criminal proceedings in which the standard for conviction is proof beyond a reasonable doubt that the electronic document was actually submitted by the signatory and that the data it contains was not submitted in error.

To satisfy this general criterion, an electronic document receiving system must establish: (1) That an electronic document was sent (or not sent), (2) when the document was sent, (3) by whom the document was sent, including both individual and the identity of any entity the individual is authorized to represent, (4) when the document was received, (5) that the document was not altered from the time it was sent to the time it was received, and (6) the contents of the document sent. In addition the electronic document receiving system must store and be able to retrieve every electronic document without alteration to its content or loss or the information regarding time of transmission, receipt, and authorship. The remaining, more specific criteria have been developed to meet these goals, while at the same time taking account of what can reasonably be expected of the various types of electronic reporting technologies currently available. It should be noted that many of these criteria will not apply, or not apply in full, where the electronic document receiving system will not be used to receive documents bearing signatures or documents used in litigation or enforcement proceedings. Generally, documents not requiring signature are less likely to play a role in criminal prosecutions; therefore, the criterion that refers to ``Validity of Data'' might not apply to systems that receive such documents. In addition, the specifications of ``electronic signature method,'' and ``electronic signature/certification scenario'' will be inapplicable, along with any provision connected with ``system security requirements,'' ``registration process,'' ``transaction record,'' and ``system archives'' that refers to signature. EPA invites comment on the exclusion of these criteria in cases where systems will not receive signed documents or documents used in litigation or enforcement and criminal proceedings. EPA will consider the possibility of developing a set of criteria explicitly addressing electronic document receiving systems that will not receive electronically signed documents if it appears that States, tribes or local governments want to implement such systems for their authorized environmental programs. Such systems might be appropriate, for example, in the cases where agencies wished to accept electronic submissions of data but continued to require that associated certification statements be signed and submitted on paper. EPA invites comment on whether it would be worth developing the alternative set of criteria for systems that exclude electronic signatures.

1. General System-Security Requirements.

Proposed section 3.2000(a) requires every system used to receive electronic documents to (1) have robust protections against unauthorized access to the system; (2) have robust protections against the unauthorized use of any electronic signature on documents received; (3) provide for the detection of unauthorized access or attempted access to the system and unauthorized use or attempted use of any electronic signature on documents received; 4) provide safeguards to prevent the modification of an electronic report once an electronic signature has been affixed; (5) ensure that every electronic record is protected from modification or deletion; (6) provide safeguards to ensure that the system clock is accurate and protected from tampering or other compromise; and (7) provide safeguards to prevent any other corruption or compromise of the system. We believe each of the seven proposed requirements is important to maintain the overall security of an electronic document receiving system. We seek comment on whether--taken together--they are sufficient to ensure that the system can maintain the integrity and authenticity of the electronic documents it receives and maintains.

2. Electronic Signature Method.

To support the goals articulated under proposed section 3.2000(b) as the "Validity of Data" criterion, proposed section 3.2000(c) stipulates that an electronic document receiving system must validate only those electronic signatures that are created by a method that (1) Involves a registration process that identifies the bearer of an electronic signature; (2) includes all elements of an adequate signature/certification scenario (described in paragraph 4, below); (3) provides safeguards to prevent excise, modification, or appropriation of an affixed electronic signature; (4) provides safeguards to prevent use of an electronic signature by anyone other than the individual to whom it has been issued; and (5) ensures that it is impossible to modify an electronic document without detection once the electronic signature has been affixed. This last proposed requirement is sometimes expressed by saying that the signature must be "bound" to the contents of the report. We seek comment on whether these conditions are appropriate, and whether--taken together--they suffice to ensure that electronic signatures affixed to electronic documents will have the same or better evidentiary value as handwritten signatures on paper documents for purposes of prosecuting an environmental crime or civil violation.

3. Submitter Registration Process.

In order to link a digital signature to the bearer of that signature, proposed section 3.2000(d) requires that an electronic document receiving system validate only those electronic signatures that are established through a process which registers identified individuals both as system users and as signature holders. EPA also proposes to require that an individual may not complete this registration process without first executing an agreement with the administering agency to properly use and protect the electronic signature.

Of course, the registration process must also establish the identity of the registering individual and any entity that the individual is authorized to represent. Given the general "Validity of Data" criterion under section 3.2000(b), the process must establish the registrant's identity with information that will be sufficient to prove that this individual was the signature holder for purposes of private litigation, enforcement proceedings, and criminal proceedings. This requires at least that the registrant provide evidence of identity which can be verified by information sources that are independent of this individual and the regulated entity with which he or she is associated.

As noted above, the rule requires that a registrant sign an agreement to properly use and protect his or her electronic signature. EPA proposes that the terms in any such agreement include, at a minimum, a commitment to: (1) Protect the electronic signature from unauthorized use; (2) be as legally-bound by use of the electronic signature as by hand-written signature; (3) where the signature device is based on a secret, e.g., a code, to maintain the secrecy of the electronic signature device; (4) immediately report any evidence that the electronic signature has been compromised; and (5) where the assistance of third parties may be required to protect a signature from unauthorized use-- such as the assistance of system administrators in ensuring computer security, to secure such assistance. EPA believes that this agreement is important to ensure that the holder of an electronic signature understands how to properly use and protect the electronic signature. It is also important to ensure that the signature holder understand the legal effect of affixing the electronic signature to an electronic document. A proof that an individual's registered electronic signature was affixed to a document will establish a permissive inference that the individual who was issued that signature affixed the signature and did so with the intent to sign the document. To achieve these goals, EPA believes that the signature agreement should consist of at least the following language:

> "In accepting the electronic signature issued by [specify name of issuing agency or organization] to sign electronic documents submitted to [specify the name of the electronic document receiving system] on behalf of [specify the name of regulated entity the signature-holder represents], I, [name of electronic signature holder], (1) Agree to protect the signature from use by anyone except me, and to confirm system security with third parties where necessary. Specifically, I agree to [specify procedures appropriate to the form of electronic signature, for example, to maintain the secrecy of the code where the signature is based on a secret code]; (2) Understand and agree that I will be held as legally bound, obligated, or responsible by my use of my electronic signature as I would be using my hand-written signature, and that legal action can be taken against me based on my use of my electronic signature in submitting an electronic document to [specify the name of the receiving agency]; (3) Agree never to delegate the use of my electronic signature or make my signature available for use by anyone else; (4) Understand that whenever I electronically sign and submit an electronic document to [specify the name of the electronic document receiving system], acknowledgments and a copy of my submission as received will be made available to me; (5) Agree to review the acknowledgments and copies of documents I electronically sign and submit to [specify the name of the electronic

document receiving system]; (6) Agree to report to [specify the agency or organization to be reported to], within twenty-four (24) hours of discovery, any evidence of the loss, theft, or other compromise of any component of my electronic signature; (7) Agree to report to [specify the agency or organization to be reported to], within twenty-four (24) hours of discovery, any evidence of discrepancy between an electronic document I have signed and submitted and what [specify the name of the electronic document receiving system] has received from me; (8) Agree to notify [specify the agency or organization to be reported to] if I cease to represent [specify the name of regulated entity the signature-holder represents] as signatory of that organization's electronic submissions to [specify the name of the electronic document receiving system] as soon as this change in relationship occurs and to sign a surrender certification at that time."

In addition, given the importance of this agreement, EPA is also proposing that the registration process require that the agreement be renewed periodically, with the Administrator to determine the frequency of and the exact terms of the renewal statement, as well as whether a wet ink signature will be required. In making these determinations, EPA is proposing that the Administrator ensure that electronic reporting meets the overall goals of security and validity of data--articulated under proposed sections 3.2000(a) and 3.2000(b)--while taking into account the importance of keeping EPA practices consistent with marketplace standards for issuance and use of electronic signature devices in commerce. Given that both the technologies and marketplace practices surrounding electronic signatures are still evolving rapidly, EPA believes that the Administrator may need to revisit these determinations more than once, the proposed provision for these renewal agreements is intended to provide this flexibility. In terms of frequency of renewal, likely candidates for the Administrator to consider are once every two years or three years, but he or she may certainly set a longer renewal cycle (either in general or with regard to a particular State, tribal or local government system) if less frequent renewal better corresponds to marketplace standards and can be determined to still meet security and validity of data goals. EPA seeks comment on the various alternatives for renewal frequency--including one year and longer than three years--considering both marketplace standards and the goals of security and validity of data. EPA also seeks comment on whether any of the candidate renewal cycles would raise any administrative issues for State, tribal or local governments, and whether the Administrator's ability to revisit this determination--with the implied potential for a change in system requirements--poses any problems for systems planning or management.

Concerning the terms of the renewal agreement, EPA believes that in the interest of supporting the goals of security and validity of data, the Administrator is likely to require the holder of the electronic signature to attest to compliance with the terms of the prior agreement since the time it was signed. To accomplish this, the Administrator may require that the signature-holder sign a statement that consists of at least the following:

"In continuing to use the electronic signature issued by [specify name of issuing agency or organization] to sign electronic documents submitted to [specify the name of the electronic document receiving system] on behalf of [specify the name of regulated entity the signature-holder represents], I, [name of electronic signature holder] continue to, (1) Agree to protect the signature from use by anyone except me, specifically, to [specify procedures appropriate to the form of electronic signature, for example, to maintain the secrecy of the code where the signature is based on a secret code]; (2) Understand and agree that I will be held as legally bound, obligated, or responsible by my use of my electronic signature as I would be by using my hand-written signature, and that legal action can be taken against me based on my use of my electronic signature in submitting an electronic document to [specify the name of the receiving agency]; (3) Agree never to delegate the use of my electronic signature or make my signature available for use by anyone else; (4) Understand that whenever I electronically sign and submit an electronic document to [specify the name of the electronic document receiving system], acknowledgments and a copy of my submission as received will be made available to me; (5) Agree to review the acknowledgments and copies of documents I electronically sign and submit to [specify the name of the electronic document receiving system]; (6) Agree to report to [specify the agency or organization to be reported to], within twenty-four (24) hours of discovery, any evidence of the loss, theft, or other compromise of any component of my electronic signature; (7) Agree to report to [specify the agency or organization to be reported to], within twenty-four (24) hours of discovery, any evidence of discrepancy between an electronic document I have signed and submitted and what [specify the name of the electronic document receiving system] has received from me; (8) Agree to notify [specify the agency or organization to be reported to] if I cease to represent [specify the name of regulated entity the signature-holder represents] as signatory of that organization's electronic submissions to [specify the name of the electronic document receiving system] as soon as this change in relationship occurs and to sign a surrender certification at that time. Moreover, I certify that I have complied with the terms of the signature registration agreement I signed on [insert date of prior agreement], and since that date I have reviewed, signed and submitted all the electronic documents submitted with my electronic signature to [specify the name of the electronic document receiving system] on behalf of [specify the name of regulated entity the signature-holder represents]."

EPA seeks comment on all of these proposed registration agreement and renewal statement provisions, including the proposed provision for administrative determination of the frequency and terms of the renewal agreements. Given the purpose of these agreements and renewal statements, EPA is particularly interested in comment on whether all of them are necessary, particularly considering requirements for the on-screen certification described under Electronic Signature/Certification, in the next section of this preamble (Section IV.D.4). To the extent that all these agreements and renewals are necessary, EPA also seeks comment on whether the specific language suggested for each provision is adequate or necessary. It should be noted that EPA is currently not proposing to codify the specific language for these certifications and statements in the rule, and EPA seeks comments on the question of codification. It should also be noted that the proposed rule specifies that the signature agreement be signed on paper or in other media that EPA may designate. While EPA will initially require signature agreements to be signed on paper--and the Administrator may initially require this of renewals as well--EPA has the flexibility to allow electronic signatures in the future, as circumstances may warrant, and when EPA believes that electronic signatures can effectively substitute for hand-written signatures on paper for these electronic signature agreements and renewals. EPA seeks comment on whether any or all of these agreements and statements should be signed on paper.

EPA also seeks comment on a possible additional certification statement, required to be signed when a signature holder surrenders the signature for whatever reason--e.g., change of jobs or retirement--although this requirement is not included as a provision in today's proposal. In this surrender certification, the signature holder would be required to truthfully attest to compliance with the terms of the agreement since the most recent agreement was signed. If such a requirement is added, then EPA believes that the surrender certification signed by the signature holder should consist of at least the following:

"I certify that, since the time that I was first issued the electronic signature by [specify name of issuing agency or organization] to sign electronic documents submitted to [specify the name of the electronic document receiving system] on behalf of [specify the name of regulated entity the signature-holder represents], I have complied with the terms of agreement to which I then subscribed, and specifically that I have: (1) Protected the signature from use by anyone except me. Specifically, I have [specify procedures appropriate to the form of electronic signature, for example, maintained the secrecy of the code where the signature is based on a secret code]; (2) Understood that I am held as legally bound, obligated, or responsible by my use of my electronic signature as I would be using my hand-written signature and that legal action can be taken against me based on my use of my electronic signature in submitting an electronic document to [specify the name of the receiving agency]; (3) Never delegated the use of my electronic signature or made my signature available for use by anyone else; (4) Understood that whenever I electronically signed and submitted an electronic document to [specify the name of the electronic document receiving system], acknowledgments and a copy of my submission as received were made available to me; (5) Reviewed the acknowledgments and copies of documents I electronically signed and submitted to [specify the name of the electronic document receiving system]; (6) Reported to [specify the agency or organization to be reported to], within twenty-four (24) hours of discovery, if I ever had any evidence of the loss, theft, or other compromise of any component of my electronic signature; (7) Reported to [specify the agency or organization to be reported to], within twenty-four (24) hours of discovery, if I ever had any evidence of discrepancy between an electronic document I signed and submitted and what [specify the name of the electronic document receiving system] had received from me. Moreover, I certify that I have complied with the terms of the signature registration agreement I signed on [insert date of the agreement signed when electronic signature was first issued], and since that date I have reviewed, signed and submitted all the electronic documents submitted with my electronic signature to [specify the name of the electronic document receiving system] on behalf of [specify the name of regulated entity the signature-holder represents]."

Finally, EPA also solicits comment on whether some other mechanism is needed, in lieu of the registration agreement, to ensure that holders of electronic signatures properly use and protect their signatures. Specifically, EPA seeks comment on the possible alternative of adding a provision paralleling 21 CFR section 11.100(c)(2) (under the Food and Drug Administration's electronic signature rule) requiring that signature holders, upon request, "provide additional certification or testimony that a specific electronic signature is the legally binding equivalent of the signer's handwritten signature." EPA seeks comment on whether codifying such a provision would provide a better method of ensuring the proper use and protection of signatures than the agreements, renewals and related certification statements that we are currently proposing.

EPA also proposes to require that an electronic document receiving system have a mechanism to automatically revoke an electronic signature whenever 1) there is any evidence the submitter has violated the registration agreement; 2) there is any evidence the electronic signature has been compromised; or 3) there is notification from an entity that the holder of an electronic signature previously authorized to represent that entity is no longer authorized to represent the entity. Revocation of a signature would not necessarily mean that the signature holder cannot be held accountable for previous uses of that signature, but it might lead the agency involved to require that particular materials be resubmitted. EPA seeks comment on whether there are other circumstances that should result in automatic invalidation of an electronic signature.

It should be added that EPA proposes to require registration of any individual who submits electronic documents to an electronic document receiving system on behalf of an entity, regardless of whether the individual is issued an electronic signature, because EPA believes that registration strengthens system security and data integrity. Accordingly, the registration process for an individual who is not being issued an electronic signature will simply omit the signature-specific requirements. EPA seeks comment on this more general registration requirement.

4. Electronic Signature/Certification Scenario.

In order for electronic document receiving systems to provide the same functionality as existing paper-based systems, the act of affixing an electronic signature to an electronic document must have the same meaning and legal effect as signing a paper document. In some instances, a signature indicates an intent to be bound to the commitments made in a document and constitutes an assertion that contents of the document are both truthful and accurate. In order to ensure that an electronic signature has the same meaning as its handwritten, paper counterpart, proposed section 3.2000(e) would require that an electronic document receiving system validate only those electronic signatures that are generated or affixed to an electronic document using a "signature/certification scenario" that ensures that the signatory understands and intends the legal consequence of affixing an electronic signature to an electronic document. This feature of an electronic document receiving system is important to ensure that each signed electronic document it receives can be used in civil and criminal enforcement, including cases against the holder of the electronic signature as signer of the electronic document.

EPA proposes to require than an electronic document receiving system must validate only electronic signatures that have been affixed after: (1) The submitter has scrolled through on-screen pages that present all the data to be certified in a familiar, human-readable format (Sec. 3.2000(e)(1)(i)); (2) the screen displays a certification statement that is similar or identical to the certifying language required on the corresponding paper submissions of the report, this display occurring just above the place on the screen where the submitter is prompted to initiate the signing process (Sec. 3.2000(e)(1)(ii)); and (3) the submitter has seen a warning--prominently displayed together with the certification statement described in (2)--that by initiating the signing process the submitter agrees that he or she is using the signature in compliance with the signature agreement that was signed when the signature device was issued (Sec. 3.2000(e)(1)(ii)).

The point of the first proposed condition is to ensure that the submitter reviews that data being submitted as a part of the signing process. Accordingly, an acceptable system must display the data in a format that clearly associates each data element with the name or label of the corresponding data field and also allow the submitter to carefully review all the data without time constraint. The point of the third proposed condition is to make certain the submitter fully understands that by activating the signature, he or she is taking a step with the same legal implications as signing and sending a report on paper. EPA is proposing this condition because of many environmental programs under which signing and certifying a false report--whether on paper or electronically--may subject the signatory to criminal prosecution. At least for those cases where the "click of a mouse" may create the potential for criminal liability, then, EPA believes it is important to ensure that the submitter understands what the consequences of the act might be. For this purpose, EPA believes that this warning statement should consist of at least the following:

> "WARNING: By signing this report, you agree that you are [name of authorized signature holder], have protected the security of your electronic signature as required by the electronic signature agreement which you signed on [date of most recent signing], and are otherwise using your electronic signature in accordance with that agreement."

Although we are not proposing to codify this language in the rule. EPA seeks comments on whether this language should be codified, and, more generally, on whether the three conditions to be satisfied prior to signing are necessary and sufficient to establish that the signature was affixed with the requisite intent.

EPA also seeks comment on three alternative versions of this third proposed condition that would replace the ``together with a prominently displayed warning. * * *." language of (Sec. 3.2000(e)(1)(ii)) with a separate provision to be inserted just before (Sec. 3.2000(e)(1)(ii)). The simplest version would read:

"The signatory attests to compliance with an electronic signature agreement that is presented on-screen, refers to the signatory by name, and includes an acknowledgment that the signatory is the authorized registrant to whom the signature was issued; and * * *".

A more robust version would read:

"The signatory attests to a statement that he or she is the authorized registrant--referred to by name--to whom the signature was issued, has taken reasonable steps to protect the signature, and does not have any reason to think that the signature has been used by anyone else; and ***".

The most robust version would read:

"The signatory attests to compliance with an electronic signatureagreement that is presented on-screen, refers to the signatory by name, and includes an acknowledgment that the signatory is the authorized registrant to whom the signature was issued, has not in the past authorized any other person to sign on his or her behalf, has not at any time compromised the electronic signature, has reviewed all automatic acknowledgments for past submissions as described in paragraph (e)(2) of this section, and has no evidence that the signatory's electronic signature or any other feature of the electronic submission mechanism has been compromised; and * * *".

Corresponding to the three versions of the proposed regulatory provision, the suggested (but not proposed to be codified) language would be, starting with the simplest:

(1) "I, [name of signatory], am the authorized holder of the electronic signature I am about to use; or

(2) "I understand and agree that I will be held as legally bound, obligated, or responsible by my use of my electronic signature as I would by using my hand-written signature."

Next, the more robust:

(1) "I, [name of signatory], am the authorized holder of the electronic signature I am about to use;
(2) "I have taken reasonable steps to protect my signature; or
(3) "To the best of my knowledge, my signature has never been used by anyone else."

And, finally, the most robust:

(1) "I, [name of signatory], am the authorized holder of the electronic signature I am about to use;
(2) "I have taken reasonable steps to protect my signature;
(3) "To the best of my knowledge, my signature has never been used by anyone else;
(4) "I have no other evidence that any component of my electronic signature has been lost, stolen or compromised in any way; or
(5) "I have reviewed all the acknowledgments and copies of my previous submissions to [specify the name of the electronic document receiving system]."

EPA seeks comment on the appropriateness of these variant alternatives to the proposed 'warning' provision--and their corresponding suggested statements--for purposes of establishing the intent with which the signature was applied, helping to show that the signatory was in fact the authorized signature holder, and preventing signature compromise or repudiation. EPA is especially interested in the question of whether any of these provisions might tend to discourage regulated entities from choosing to submit environmental reports electronically. EPA is also interested in comments on the need for any version of this 'warning' provision in view of the certifications provided in conjunction with the renewals of signature agreement discussed in the preceding section of this preamble (Section IV.D.3).

In addition, we are proposing that, once the electronic signature is affixed, and the electronic document submitted, the signature/certification scenario must include two responses from the electronic document receiving system. The first is simply an automatic acknowledgment that the report has been received and any affixed electronic signature validated, with the time and date of receipt. The purpose of this acknowledgment is, at least in part, to alert the registered holder of an electronic signature if someone has appropriated the registered electronic signature and used it to submit spurious electronic documents. As noted above, the registered holder of the electronic signature will not be allowed to sign another electronic document once aware that it has been compromised.

EPA also proposes to require that the automatic acknowledgment be sent to an address that does not share the same access control--for example, that is not protected by the same passwords or confidential log-in procedures--as the system from which the electronic report was signed and sent. The intent of this requirement is to frustrate unauthorized use of an electronic signature without detection. To elude detection, the intruder will have to compromise not only the signature protections, but also the additional system's access controls. The additional address could be electronic or could be a United States Postal Service address. In any event, the feature of the electronic document receiving system should aid in the detection of compromised electronic signatures and reduce the requency and strength of false claims that an electronic signature has been appropriated without the knowledge of the registered holder of the electronic signature. The second response is what we are calling the 'copy of record', also automatically created and made available to the submitter. The copy of record must include the complete electronic document that was submitted. The copy of record must be complete in the sense that it must accurately associate all of the information provided by the submitter with the descriptions or labeling of the information being requested. In addition, to be complete, the copy of record must include all the warnings, instructions and certification statements presented to the submitter as a part of the signature/certification scenario. Finally, this copy of record must: (1) Be viewable on-screen in a human-readable format that makes clear the association between each of the information elements provided by the submitter and the descriptions or labels in terms of which these elements were requested; (2) include the date and time of receipt; and (3) be signed with a secure, immutable agency electronic signature that is ``bound" to this electronic document. As the name would suggest, the copy of record must be archived by the agency system, made available to the submitter for viewing and downloading, and protected from unauthorized access.

The proposed copy of record requirement is intended to detect spurious or compromised submissions, enabling timely disavowal of unintended submissions and reducing the frequency and strength of claims that an electronic document has been modified in transmission or unintentionally submitted. Under the signature/certification scenario in today's proposed rule, the copy of record will be--strictly speaking--made available to the registered holder of the electronic signature. If the signature has somehow been compromised--or if the data is somehow different from what was intended to be submitted--this copy of record, together with the acknowledgments discussed above, will give the signature-holder an opportunity to alert the agency to the compromise of his/her signature and/or his/her data. This proposed requirement is also intended to protect the agency from attempts to falsely repudiate a submission.

EPA seeks comment on whether the number and type of responses from the electronic document receiving system adequately address the issue of spurious or compromised submissions. Specifically, we seek comment on the requirements placed on the automatic acknowledgments. In addition, we are interested in views on whether it will be generally feasible for electronic document receiving systems to create copies of record with all the attributes we are proposing that they have, and whether all of these attributes are necessary for the copy of record to fulfill its intended purpose.

5. Transaction Record.

To help settle potential disputes over whether certain submissions were made, when they were made, what they contained, or who made them, an electronic document receiving system must create a transaction record for every submission of an electronic document. EPA will require that this record be created automatically, and include the precise routing of the signed electronic document from the submitter's computer to the receiving system and the copy of record described above. In addition, based on the receiving system's clock, this transaction record must include the precise date and time of: (1) The initial receipt of the reported data; (2) the receipt of the submitter's signed certification of the data (where this step is subsequent to the initial data transfer); (3) the sending of the acknowledgment notice; and (4) the creation of the copy of record. These details may be regarded as providing the "chain of custody" for the submitted report, and help to establish its authenticity. EPA seeks comment on whether this transaction record specification is sufficiently robust to provide for "chain of custody".

6. System Archives.

EPA also proposes to require that electronic document receiving systems maintain the contents of the transaction record described above--including the copy of record--for as long as they may be needed for enforcement or other programmatic purposes. In addition we are also proposing that the system must maintain records that show, for any given electronic submission not only what information was displayed to the user during the submission process--including the instructions, prompts, data labels, etc. captured in the copy of record--but also how this information was displayed, including the sequencing, functioning and overall appearance of these interface elements. The reason is that it may be difficult to interpret what some of the submission's data elements mean if we do not know the context within which they were provided--e.g., to what on-screen display or query a "yes" was responding. Depending on exactly how the signing process is implemented, at least some of this interface information may be captured within the scope of what is bound by the signature, e.g., if the signature is applied to the entire content of the screens that are reviewed by the signatory during the signature/certification scenario. To whatever extent this occurs, the archiving of the "copy of record" would contribute to this archiving of the interface.

The system must maintain the archived records in a way that can be shown to have preserved them without any modification since the time they were created; the system must be able to make these records available to users in a timely way as they are needed. EPA seeks comments on these archiving criteria, and especially on whether there are any issues raised by the need to maintain the copy of record—which includes electronic signatures--over long periods of time. Of particular concern are copies of record that include digital signatures, as they will for electronic submissions received by the Central Data Exchange (CDX). (For an explanation of digital signatures, and their role in CDX, see Section V.B.1 of this preamble.) Ideally, the system will preserve digital signatures in a form which allows them to be validated at any point during the life of the archived records that contain them; this is the standard implied by Sec. 3.2000(g)(2)(i) that requires the copies of record to be preserved "in their entirety" for the life of the archive. However, EPA realizes that this ideal may be difficult to implement in practice for several reasons, including:

- The sensitivity of digital signatures to very minimal (and unavoidable) deterioration of the magnetic medium in which the records are stored--so that they no longer can be validated, even though the records remain usable in every other way;
- The possible software dependence of the validation process--so that, as the archives' systems environment evolves over long periods of time, it may become increasingly difficult to operate the validation software designed to work with the archived signatures; and
- The dependence of validation on the accessibility of a public key infrastructure (PKI) certificate that was valid when the digital signature was created--so that, over time, it may become increasingly difficult to determine the keys and identifying information associated with the signature.

EPA seeks comments on these and related difficulties that may stand in the way of validating archived digital signatures, and we welcome any advice on how these might be overcome. If these difficulties cannot be overcome, or overcome only at great expense, then EPA would seek to revise Sec. 3.2000(g)(2), by specifying alternatives to maintenance of the original signature and its validation as archived that would still allow users to demonstrate both the validity of the signature and the integrity of the record as a true picture of the data as it was signed. A possible approach might involve an archivists' wet-ink-on-paper certification that the digital signature was valid at the time the record was placed in the archive, together with appropriate measures to preserve the record unchanged. On another approach, the archivist might digitally resign the document at certain intervals, adding appropriate certifications about the validity of the original (or previous) signature on the document. EPA also seeks comment on such alternative approaches.

E. What Are the Costs and Benefits Associated With Today's Proposal?

EPA estimates that today's proposal could result in an average annual reduction in reporting and record-keeping costs for those information collections identified as potentially benefitting from offering an electronic reporting option. Based on this analysis, EPA estimates that CROMERRR could result in an average annual reduction in burden of $52.3 million per year for those facilities reporting, $1.2 million per year for EPA, and $1.24 million for each of the 30 states that were assumed to implement programs over the eight years of the analysis. For details of this study, see the technical background document, Cross Media Electronic Reporting and Recordkeeping Rule Cost Benefit Analysis in the Docket for today's proposal. EPA requests comment on whether the underlying assumptions and the methods used in the cost benefit analysis provide a realistic estimate of the costs and benefits associated with electronic reporting and recordkeeping.

1. Scope and Method.

The purposes of the analysis was to estimate the labor hour and total cost effects (either savings or increases) attributable to each of the major elements of the CROMERRR proposal and to assess, qualitatively, the environmental implications. The major elements include: the use of modern electronic technologies for the production, completion, signing, transmitting, and recording without the use of paper copies. Within the assessment of technologies we chose three forms of electronic reporting (web forms, EDI, and XML) that EPA's CDX plans to support. For those entities using web forms, the costs of reporting to EPA electronically would be negligible, as EPA intends to provide the web forms and signature capabilities needed. In the latter two approaches (EDI and XML), EPA anticipates additional up-front cost will be incurred by regulated entities to establish EDI or XML file generation capabilities, but the savings will be larger over time, as these entities can more fully automate their reporting to EPA. In the course of establishing projected estimates of costs and savings of electronic reporting and recordkeeping, EPA had to establish a baseline of current costs. The current costs of paper-based reporting to EPA and States delegated the authority to manage an EPA reporting program were based on an extensive assessment of EPA's official information collection request (ICR) submissions that would be subject to the CROMERRR rule, as well as more detailed cost estimates performed on major EPA systems. In performing the analysis, over 50 ICRs were extensively reviewed and approximately 70 other ICRs were more summarily reviewed. A list of the ICRs, and the approach used to analyze them, are contained in Appendix A of EPA's Cross Media Electronic Reporting and Recordkeeping Rule Cost Benefit Analysis. In the course of analyzing the ICR costs, reporting costs were broken into discrete functional areas (such as data entry, mailing, reconciliation, archiving and program management) and were analyzed for costs. In addition to the ICR analysis, EPA performed analysis of the general costs and benefits of electronic reporting experienced by commercial and government agencies, as described in the EPA Electronic Reporting Benefit/Cost Justification Report (June 30, 1999). EPA also conducted in-depth analyses of business processes and associated costs for several major EPA programs. These analyses include analyses for Toxic Release Inventory (TRI), National Pollutant Discharge Elimination System (NPDES), Public Water Supply System (PWSS) and selected Clean Air Act reports. In addition, EPA, in conjunction with State partners in the Arizona Department of Environmental Quality (ADEQ) and the Texas Natural Resources Conservation Commission (TNRCC), conducted assessments of the potential impacts and opportunities presented by environmental electronic reporting on their EPA-delegated state programs and affected regulated entities. These programmatic and state analyses are available in the CROMERRR docket. EPA also reviewed similar analyses performed for other EPA electronic reporting efforts, such as the proposed Hazardous Waste Manifest Automation Rule. EPA invites comments on the approach used for conducting the analysis and on the list of ICRs analyzed--whether this list encompasses the spectrum of EPA requirements impacted by CROMERRR and what additional information collections, if any, should be incorporated into further analysis.

Based on the combined review of the functional areas (including data entry, mailing, reconciliation, archiving and program management) of individual ICRs, EPA identified general trends in the relative distribution of costs for each of the categories. Using the analyses conducted under the more in-depth studies performed, EPA was able to estimate the impacts of electronic reporting on each of the functional areas (including data entry, mailing, reconciliation, archiving and program management). For instance, by offering facilities the electronic submission as an alternative to printing and mailing the paper submissions, the percentage of costs attributed to "mailing" could be eliminated. Using this logic, EPA added the relative percentages of reductions in each of these functional areas, and determined that a general reduction of 11 percent in the overall cost of reporting could be achieved through web-based submissions, and that a 25 percent reduction could be achieved for those facilities that implement EDI or XML based exchanges.

EPA is also considering a second series of analyses, using an alternative form of calculating the costs and savings to the Agency. In performing this alternative analysis EPA would still break the costs for a program report into discrete functional areas (i.e., data entry, mailing, etc.), however the estimates of reduction would use "absolute" values instead of percentages. As an example, EPA program X has identified that the mailing of form B requires 10 minutes per submission. The costs for facilities choosing to submit electronically would take into account the elimination of mailing, and the costs for electronic reporting under that program would be reduced by 10 minutes for each submission. The advantage of this approach is that it offers potentially greater accuracy for estimating costs for each reporting program. A disadvantage is where the functional activity, such as program management, is only partially impacted by electronic reporting, determining an "absolute" value could involve arbitrary judgement calls on a program by program basis. EPA requests comment on ways to improve an analysis of this type as well as suggestions for other approaches that may better identify the potential costs and benefits of the proposed electronic reporting and recordkeeping rule.

As discussed further below, two sets of regulatory cost reduction (savings) estimates were projected--one for web based submissions and one for EDI/XML--based on a range of alternate assumptions regarding the national adoption rates for automation options. In both cases, it was assumed that 77 percent of all reports would be prepared, transmitted, and recorded electronically at full implementation. The implementation rates of facilities, however, will vary depending on the degree to which the facility implements electronic reporting for environmental requirements directly with EPA or with State regulatory agencies managing EPA-delegated/authorized environmental programs. The rates are also affected by the method (Web, EDI, or XML) the facility chooses to use in reporting to EPA or the delegated State agency. Table III describes the implementation rates for facilities under the scenarios described. The Table also presents the current "As-Is" rates of paper or diskette exchange and the impacts of electronic reporting on these rates over an eight year period. Recordkeeping rates are not presented in this Table. However, it was also assumed that a very low number of facilities (0.5 percent) of the current regulated entities, would elect to acquire new electronic recordkeeping systems to implement the CROMERRR recordkeeping option. EPA is seeking comments on the implementation rates for reporting and recordkeeping as described in this proposed rule.

Table III. Facility Implementation Rates by Reporting Method [In Percent]

Reporting method	FY00	FY01	FY02	FY03	FY04	FY05	FY06	FY07
As-is:								
Delegated	100	100	95	89	81	73	64	56
Non-delegated	100	100	96	66	50	45	36	28
Mixed delegation	100	100	96	77	66	59	50	42
Web								
Delegated	0	0	4	8	12	18	24	30
Non-delegated	0	0	3	25	32	37	42	48
Mixed delegation	0	0	3	17	22	27	33	39
EDI								
Delegated	0	0	1	2	2	3	4	5
Non-delegated	0	0	1	4	6	6	7	8
Mixed delegation	0	0	1	3	4	5	6	6
XML:								
Delegated	0	0	0	2	4	6	8	10
Non-delegated	0	0	0	4	12	12	14	16
Mixed delegation	0	0	0	3	8	9	11	13

For EPA, the average annual cost to implement and operate electronic reporting and record-keeping is $25.8 million, and the average annual cost savings compared to equivalent paper-based systems is $1.2 million. The average annual cost to implement an electronic reporting system is $1.1 million for each state, and $1,273 for each facility. The net average annual cost savings of electronic reporting compared to an equivalent paper-based submission is $1.24 million for each state, and $1,140 for each facility. The total average annual costs of implementing and reporting electronically for all facilities is $3,420 million, which presents a net average annual savings for all facilities of $52.3 million over current paper-based reporting. The average annual cost to implement a new electronic record keeping system is $40,000 for each facility, and the net average annual cost savings for operating the electronic record keeping system is $23,080. These costs are based on FY 2000 dollars and include a 7.0 % annual discount rate. Therefore, our estimates indicate that implementation of electronic reporting will result in a net burden reduction for all participants, but facilities may not find it cost-effective to develop an electronic records system unless it addresses both EPA and non-EPA business purposes. Table IV summarizes the total cost of the current "as is" paper system and the future "to be" electronic reporting and record-keeping costs over the next eight (8) years for EPA, States, and regulated entities. In preparing this analysis, EPA chose to be conservative in assigning implementation rates and used technology costs based on the current year.

It should be stressed that the facility cost and cost-savings estimates that these totals represent are averages per facility, and these averages cannot be translated into costs/cost-savings per report submitted electronically. The cost-related effects of introducing electronic reporting for a particular report may depend on circumstances that are unique to the data being reported, and these specifics are not reflected in the per facility averages. Accordingly, while the facility cost and cost-savings estimates are based in part on considering the ICRs that are likely to be affected by the proposed rule, the resulting cost/cost-savings numbers cannot be used 'in reverse' to calculate cost and burden reductions associated with introducing electronic reporting for any individual ICR.

In addition, the actual costs and cost-savings for implementing facilities will vary widely depending on the electronic submission approach. Companies choosing to submit using web forms will have much lower initial investment costs, but will receive less savings than companies that choose to automate their systems to generate EDI or XML file submissions to EPA. In the latter case, EPA assumes that costs associated with the implementation of EDI or XML will result from companies configuring existing XML or EDI software to EPA prescribed formats, and companies will tend not to invest in EDI hardware or software for the singular purpose of submitting data to EPA. If the electronic

Table IV. Summary As-Is Versus To-Be Costs and Cumulative Savings ($M) [In FY 2000 Dollars]

Cost	FY00	FY01	FY02	FY03	FY04	FY05	FY06	FY07
As-Is costs:								
Facilities	3,863.0	3,883.7	3,775.0	3,669.2	3,566.1	3,444.1	3,369.2	3,274.7
States	58.7	59.0	57.4	55.8	54.2	52.7	51.2	49.8
EPA	25.8	26.9	26.9	27.1	27.2	27.4	27.5	27.6
To-Be costs:								
Facilities	3,863.0	3883.7	3,771.3	3,629.4	3,520.8	3,357.7	3,278.7	3,197.8
States	58.7	59.0	42.3	40.1	38.4	37.5	36.2	35.0
EPA	28.4	30.7	42.3	26.9	21.5	19.6	19.3	18.4
Difference	(2.6)	(3.9)	3.5	55.6	66.8	109.3	113.8	101.0

commerce industry trends continue, the costs of implementing technologies will decline and the number of facilities and states implementing electronic reporting will increase, thereby increasing the overall net benefits of the rule. EPA is also continuing to research electronic record-keeping options that will improve the cost effectiveness of electronic record-keeping while meeting federal enforcement requirements. EPA is seeking comment from reviewers on alternative record keeping approaches and on EPA's assumption that facilities choosing to submit data via XML or EDI to EPA will not acquire new hardware or software.

2. Qualitative Implications.

In addition to the cost savings identified through implementation of this proposal, EPA also has identified a number of qualitative benefits through implementation of an electronic system. These qualitative benefits of electronic reporting include: enhanced quality of data received and entered into our systems, faster public access to data submitted to EPA, better tracking of compliance submissions by industry and government agencies, and opportunities for re-engineering current paper processes. EPA's Cross Media Electronic Reporting and Record-keeping Rule Cost Benefit Analysis describes the qualitative aspects in more detail.

V. The Central Data Exchange (CDX)

A. What Is EPA's Concept of the CDX?

EPA's Office of Environmental Information (OEI) is currently developing the specifications for a 'central data exchange' that will serve as EPA's primary gateway for electronic documents received by EPA. As noted in section I.B of this preamble, CDX is being designed with the goal of fully satisfying the criteria that this proposal specifies for assessing State or tribal electronic document receiving systems; similarly, EPA will ensure that other systems the Administrator designates to receive electronic submissions satisfy the criteria as well. With respect to the electronic document submission process and criteria addressed by today's proposal, we intend CDX functions to include:

- Access management--allowing or denying an entity access to CDX;
- Data interchange--accepting and returning data via various of file transfer mechanisms;
- Signature/certification management--providing devices and required scenarios for individuals to sign and certify what they submit;

- Submitter and data authentication--assuring that electronic signatures are valid and data is uncorrupted;
- Transaction logging--providing date, time, and source information for data received to establish ``chain of custody'';
- Acknowledgment and provision of copy of record--providing the submitter with confirmations of the data received;
- Archiving--placing files received and transmission logs into secure, long-term storage;
- Error-checking--flagging obvious errors in documents and document transactions, including duplicate documents and unauthorized submissions;
- Translation and forwarding--converting submitted documents into formats that will load to EPA databases, and forwarding them to the appropriate systems; and
- Outreach--providing education and other customer services (such as user manuals, help desk) to CDX users.

The idea is to eventually provide--to the greatest extent possible--one way and one place for the regulated community to exchange electronic documents with EPA. States may also choose to use CDX as a gateway for electronic data submissions from their regulated community, as a cost-effective alternative to building their own system. EPA is exploring opportunities to leverage CDX resources for use by authorized/approved state programs. CDX may also provide the platform for State-EPA data exchanges that implement administrative arrangements for data sharing. However, as with the provisions of the proposed rule, the features and functions of CDX described in this Section will generally be inapplicable to these State-EPA exchanges.

With respect to EPA's electronic transactions with regulated entities, our hope is that the uniformity of process and technology that CDX provides will help both EPA and regulated entities realize economies of scale from their investments in data exchange technologies. This is not to say that use of CDX to submit electronic documents will necessarily involve substantial investment; it will require little more of a submitter than access to a computer with a browser and an Internet connection. However, for organizations that have invested heavily in the computerized management of their environmental data, CDX is also being designed to support substantial automation of the data transfer processes. In addition, EPA hopes that CDX's centralization of data exchange will eventually provide the platform for greater integration or consolidation of environmental reporting.

B. What Are the CDX Building Blocks?

To support its various functions, we are designing CDX to incorporate a number of key building blocks, including:

- Digital signatures based on public key infrastructure (PKI),
- A process for registering users and managing their access to the CDX,
- A characteristic systems architecture,
- Electronic data interchange (EDI) standards, and
- A characteristic environment in which electronic reporting transactions will be conducted.

These building blocks--as explained in detail in the following sections--are meant to ensure that CDX can perform the functions of an electronic document receiving system under the proposed rule. EPA believes that these building blocks, taken together, will satisfy the criteria in today's proposal for electronic document receiving systems, but seeks comment on this general question.

1. Public Key Infrastructure (PKI)-Based Digital Signatures.

PKI-based digital signatures are the product of two concepts: "Asymmetric" cryptography, and an institutional framework for "certifying" the identity of a signature-holder, provided by PKI. Taking these in order, "asymmetric" cryptography is based on a mathematical relationship that exists between certain pairs of numbers, for example number A and number B, such that if A is used to encrypt some message, B and only B can decipher it, and if B deciphers the message, it can only have been encrypted with A. For purposes of a digital signature, then, A and B are uniquely assigned to individual X. (How this works is described below, in connection with explaining the "institutional framework" provided by PKI.) One of the numbers, say A, submitter X shares with no-one. This is X's "private key". The other, B, is X's "public key", and X shares B with anyone to whom X wishes to send a message--X may even publish B together with information that identifies him/her as X.

Given his two keys, X then signs an electronic document as follows: (1) X uses a standard formula or algorithm to produce a number uniquely related to the content of the electronic document. This is referred to as the "message digest" or "hash" of the document; (2) X uses A, the private key, to encrypt this hash; this encrypted hash is X's digital signature, and it is unique both to X and to the particular message it signs; and (3) X attaches this digital signature to his/her message (which is otherwise not encrypted), and sends it.

When Y gets X's message, Y validates X's signature by: (1) Deriving the hash of the message, using the same standard algorithm that X used; (2) deciphering X's digital signature, using X's public key, B; and (3) comparing the hash Y derived (in step1) with the deciphered signature. The two numbers-- the derived hash and the deciphered signature--should agree. If (and only if) they do, then Y knows both that the signature was produced using A (which belongs to X), and that the message has not changed since X signed it.

Because the digital signature is specific to the particular document, and is unique in each case, to say that X is a "signature-holder" in this context is to refer to A and B, the private/public key-pair. The A/B key-pair does belong to X and plays the same role in each of the many digital signatures X may create through the process described above. Accordingly, it is this key-pair--rather than the individual signatures they are used to create--that is associated with the process of certifying a signature-holder's identity that is provided by PKI.

Turning to this, PKI is a way of reliably establishing and maintaining the identity of the individual associated with a given key-pair used in producing digital signatures. This protocol involves the issuance of a "PKI certificate" by a "trusted" "certificate authority" (CA). The CA is "trusted" in the sense that it operates in conformance with an appropriate certificate policy, and has demonstrated this conformance through its operations across a wide range of electronic commerce applications.

Issuing a certificate for individual X typically involves the following steps: (1) X applies to the CA for a certificate; (2) the CA requests various pieces of personal information from X, and/or notarized verifications of X's personal information, and/or X to appear in person, to provide the CA with the bases for "proving" X's identity; (3) the CA provides X with a way to generate his unique key pair; (4) the CA conducts the "identity proofing" process--matching what X has provided against information about X in various commercial databases, official documents, etc.; (5) when the "identify proofing" is successfully completed, the CA creates a "certificate" for X that incorporates his public key, along with various pieces of identifying information about X; (6) the CA digitally signs the certificate to certify its authenticity, and makes it available to users through directory services. Some of these steps--especially the "identity proofing" process--may vary considerably, depending on requirements for security/certainty and the policies and practices of the particular CA. In the approach that EPA is currently planning, certificate issuance will be incorporated into a broader CDX registration process. The discussion of registration in the next section will include some of the proposed specifics of "identity proofing" and related steps for CDX purposes.

The use of PKI-based digital signatures is itself supported by a very robust infrastructure of electronic commerce tools and practices, private- and public-sector policies and standards, as well as a very large and growing body of theoretical research into the mathematical foundations for this approach. Within the federal government, the importance of PKI is recognized not only by the ACES initiative (discussed below), but also by a standing ``Federal PKI Steering Committee'' with the mandate to promote and coordinate the adoption of PKI-based digital signatures for a broad range of applications across all federal agencies. In addition, federal agencies may rely on security and PKI technical requirements published in the Federal Information Processing Standards (FIPS) developed by the National Institute of Standards and Technology, available at http://csrc.nist.gov/fips/.

2. The CDX Registration Process.

Under the system EPA is designing, to submit electronic documents to EPA you must first register with CDX, and--at least at the outset--registration will be by invitation from EPA. Generally, as CDX is readied to receive a specified report, EPA will extend registration invitations to all individuals who currently submit that report to EPA on behalf of their organizations, and are identified as having this responsibility in EPA's Facility Registry System (FRS) database. If you have this responsibility but do not receive an invitation, you will have the opportunity to notify EPA and put yourself on our invitation list. However, if you submit the specified report to a State, tribal or local agency, you will not receive a CDX invitation, since your reporting transaction would be with that agency's electronic document receiving system, and not with CDX.

If you decide to accept an invitation to report electronically, you will go through a registration process that involves three steps: Invitation and verification; Certificate issuance; and Access and agreement.

Taking these in order, EPA will initiate the process by sending you a letter, through the United States Postal Service. The letter will indicate the opportunity to report electronically, provide a CDX web-site address and access code, and invite you to start the registration process by logging on to the CDX site and verifying your name, address, organizational affiliation and area of reporting responsibility as posted on that site. This verification session will conclude by providing you with the web-site address for the Certificate Authority (CA) that will take you through step 2 of the process.

Of course, you may not have the responsibilities that the CDX site indicates. That is, you may not be the individual who signs and submits the environmental reports the site specifies on behalf of your company. In that case, you will be invited to indicate the individual(s) who do(es) have these responsibilities, and that will conclude your own interaction with CDX. EPA will then update FRS, and issue new invitation letter(s) to the correct individual(s). Assuming you are the correct individual, step 1 may in some cases involve EPA asking for a letter from a responsible company official, on company letterhead, confirming that you have the responsibility to the sign and submit the environmental reports in question. Finally, as a part of step 1 you may also be prompted to nominate one or two individuals as "alternate" submitters, to receive their own invitations to register and, via step 2, to obtain their own PKI certificates. EPA is considering this provision for "alternates" so that there will always be someone at the facility available to sign electronic submissions with their own private key, in case you-- as the primary submitter-- are unavailable during a period when a document is due. EPA seeks comment on the value of the confirming letter, and of providing for these "alternates", and on whether these would impose any unacceptable costs or burdens on regulated entities.

Moving on to step 2, certificate issuance will largely be in the hands of the certificate authority (CA). EPA's current plan is to secure CA services through the General Service Administration's (GSA) Access Certificates for Electronic Services (ACES) program. Under ACES, EPA will contract with one of the ACES vendors to issue and manage certificates for individuals wishing to submit electronic reports to CDX. More information on ACES is available at the ACES website: www.gsa.gov/aces.

Assuming the ACES approach, then, issuance of your certificate will consist of a sequence of events similar to the following:

- You log onto the ACES CA's web-site, using the address provided at the end of step 1, and the access code provided in the initial invitation letter;
- You provide personal and business information that may include some of the following items--your name, home address, e-mail address, social security number, telephone number, credit card number, driver's license information, employer's address, common name of your employer, legal company name of your employer, name and telephone number of your direct manager, and name and telephone number of a human resource contact;
- During this initial ACES CA session, the CA will also enable you to generate--on your own computer--a public and private key pair, and your public key would automatically be included in your certificate request;
- The CA will use your personal and business information to conduct the identity-proofing process; this takes approximately three days;
- After the CA validates your identity, you will receive a letter via the US Postal Service notifying you that your certificate is ready; notification will include a PIN for access to the certificate retrieval website;

- You may be asked to return to the ACES CA web site to confirm the receipt of your certificate and acknowledge that you have read and agree to abide by the conditions of your new EPA-sponsored certificate; and
- You will download the certificate to your browser, the CA notifies CDX that you have received your certificate, and CDX initiates step 3.

Under the ACES approach, the personal information you supply for purposes of ``identity proofing" must include at least three items, and at least one of these must be something assigned to you based on an in-person identity verification process, e.g., a passport number or driver's license number. In addition, because your identity as an official of a regulated company is central to your relationship with EPA, the "identity proofing" performed by the CA may also include verification of your company's identity, including address, legal name, names of directors and officers, and current operating status. EPA seeks comment on any aspect of this "identity proofing" approach, and specifically on the need to have the CA collect the personal and business information listed above, as well as any comment on the ACES certificate issuance process as a whole.

It is worth stressing that the items of personal information selected for "identity proofing" will be submitted to the CA, and not to EPA, and this personal information will not be available to or maintained by EPA. However, some basic personal information--specifically, your name, your contact information (email address, phone/fax/mobile/pager numbers), your mailing address and your organizational role (e.g., consultant, environmental manager, etc.) may be submitted to (or verified as correct by) EPA as a part of step 1 of the registration process, preceding ACES certificate issuance. Step 1 may also involve EPA's collecting or verifying some of the business-related items that can also be associated ACES "identity proofing"--specifically, your employer's address, common name of your employer, legal company name of your employer, name and telephone number of your direct manager--plus, possibly, the following additional items of information: facility name and address, EPA program reporting area (e.g., Hazardous Waste, NPDES, etc.), EPA program or permit identification number, and preferred method of electronic reporting (e.g., web form, EDI, etc.). EPA seeks comment on the need to collect/verify these items of personal and business-related information as a part of step 1 of the registration process.

In step 3, CDX will create a system account for you, including a controlled-access mailbox, sending you by regular mail the password and user identification code to gain access to your account. When you initially use these to access your account, you will be instructed to download any client desktop software from CDX that may serve to support the digital signing of your electronic submissions. You will conclude the registration process by printing out and signing on paper a registration agreement included with the downloaded software. The agreement will affirm your understanding that, among other things:

- Digital signature/certification has the full legal force of a corresponding signature created with wet ink on paper;
- You must protect the access to your CDX mailbox, to your client CDX desktop, and to the private key used to create your digital signature;
- You must never delegate the use of your private key, or provide anyone else access to it in any other way; and
- You must immediately notify EPA if you have any reason to suspect that your CDX mailbox, CDX-supplied client software, or private key has been compromised

The full agreement would conform closely to the text suggested in subsection IV.D.3 of this preamble.

Upon receiving this agreement, with wet-ink-on-paper signature, CDX will recognize you as a fully-registered and authorized user. As proposed in today's rule, CDX will require a process for you to renew your registration, probably once every two years, although--corresponding to the discussion in Section IV.D.3 of this preamble--EPA seeks comment on less frequent renewals, for example, at intervals of 3, 4, or 5 years. This will include certifying that you have complied with the terms of your initial registration agreement, and, in particular, that you have not in any way compromised or delegated access to your private key, to your private CDX account, or to your CDX client software, and that you have no other evidence that any of these items have been compromised. Again, the full text of this agreement would conform closely to the text suggested for agreement renewal in Section IV.D.3 of this preamble. This certification will probably be printed out by your desktop software, require a wet-ink-on-paper signature, and be submitted through the United States Postal Service. Failure to submit this certification would terminate your access to CDX, and could lead EPA to require supplemental certification of previous submissions. The EPA is seeking comment on this proposed approach to registration renewal, the requirement that the agreement be renewed, and the frequency of the renewal. We are also seeking comment on whether it could be accomplished via an electronic submission rather than on paper.

2. The CDX Architecture.

In designing the CDX architecture, EPA has been guided by three goals:

- Flexibility in exchanging data--that is, the ability to support a number of different data exchange mechanisms, including batch file transfers in various formats, web-based file uploads, as well as on-line data entry;
- Uniformity in signing/certifying submissions--that is, providing for a uniform way for individuals to sign and certify their electronic documents, no matter how the data they contain was transferred; and
- Adequate security for all aspects of CDX operation--that is, the assurance that authorized users of CDX, including EPA, retain control over the CDX operations for which they are responsible.

The goal of flexibility arises from knowledge that the organizations that might want to submit electronic documents to CDX apply information technology to environmental management many different ways. At the one extreme may be large companies that have correspondingly large quantities of data to submit--data that they maintain in databases and would prefer transfer in as automated a mode as possible. At the other extreme are small businesses that may be equipped to enter their data into some sort of user-friendly 'smart' form--on-line or off-line--but would not otherwise computerize their environmental data. And, in the middle, are organizations that may use relatively simple database or spreadsheet tools for their environmental data, but are not prepared to automate a data transfer process. In designing CDX, EPA in trying to accommodate all of these varying levels of computerization--providing organizations with modes of data transfer that fit their capabilities while allowing them to take advantage of whatever level of data capture and automation they have already achieved.

While organizations may differ considerably in how they want and are able to transfer data, there needs to be a consistent approach for the responsible company official's review and certification--by signing--to the truth and accuracy of the data transferred. In all cases this will be accomplished by a human interaction with the medium in which the data is displayed, and some human action to create the signature in that medium. For any case that calls for a signature, CDX will always provide the same uniform set of procedures for reviewing the data and creating the signature.

The CDX will also be designed to provide the requisite system security. Obviously, the CDX must involve protection for the data that CDX receives and maintains from any unwanted intrusion or tampering. It must also protect the data as it travels from the submitter to the CDX. The system security must also include elements that ensure that the signature/certification process is not compromised. For example, CDX must provide certificate holders with a way to secure their private key and to control access to any messages that confirm or respond to submissions, so that they can be assured that no spurious transactions with CDX will be conducted using their electronic signature.

To achieve these goals, EPA is planning to base CDX implementation on client-server architecture. This means that CDX will manage the transactions with submitters through a computer operated by EPA that interacts with computers at the submitter's site. To provide for the desired flexibility, the EPA server is being designed to accept data via a variety of transfer mechanisms in variety of formats, ranging from Internet File Transfer Protocol (FTP) submissions of spread-sheet files to standards-based electronic data interchange (EDI) transmissions via private value-added network (VAN). These file formats and transfer protocols will be discussed below.

To ensure a uniform signature/certification process, CDX would provide the computers from which it accepts electronic documents (otherwise known as "client" personal computers (PCs)) with copy-protected and password-protected client software that will support the digital signing of your electronic documents. You will be prompted to download and install this software once you complete the registration/certification process, and access your password-protected mailbox on the CDX server. (You would also be given a detailed user's guide, which will provide step-by-step instructions on download and installation.).

To operate this CDX client software, and interact with the CDX server, your PC system will have to have: Internet access; at least a 486 processor (with Pentium recommended); 2 to 5 MB of available hard-drive space to install program software; access to a printer; and Microsoft Windows 95, 98 or NT 4.0. Given the planned use of digital signature certificates, your system will also be required to run one of the following Web browsers: Internet Explorer 4.01, Internet Explorer 5.0, Netscape 3-4.05, Netscape 4, or subsequent versions of these browsers. In addition, you should have backup capability of some form (e.g. tape system, off-line disk storage, or access to a separate network server.); an effective backup program provides protection against system malfunctions and ensures that you can retain a copy of your submissions as required by EPA regulations. EPA seeks comment on whether these system requirements impose unacceptable costs or burdens on regulated entities, and whether additional processors and operating systems should be accommodated.

Concerning protection of the server, CDX will be designed to incorporate "firewall" security, in addition to the usual system security provisions to control physical access to the system and prohibit unauthorized internal access. Very generally, a "firewall" is software that controls the flow of data files between a system and a network to which it is connected, to ensure (among other things) that only files from recognized and safe sources are allowed to enter. As transmissions flow through the CDX firewall, for example, they will be automatically virus-scanned, and the system would not attempt to process a file that contains a suspected virus. (If a virus is detected, the submitter would be notified and asked to resubmit the report.) The server will also be protected with intrusion detection software that alerts the system operators to suspected attempts to penetrate or "hack" the system. The system operators will use the logging capability of the firewall and the intrusion detection system to monitor the health and status of the system and respond to unauthorized efforts to use or modify the system. In terms of protecting the system clock, CDX will be configured so that changes to the clock can only be made under a single user ID and password, and the server will be placed in a locked rack so that an unauthorized person cannot use a reboot sequence to change the clock settings. In addition, the system clock will be synchronized with the atomic clock at least once a day to ensure that the system time is extremely accurate.

Once a submission passes through the firewall, CDX will initiate the first of several processes that, among other things, will create a robust archive of the original submission, including:

- The submission files in their entirety, exactly as they were sent, including any enveloping/addressing/routing/date-time information. These will be captured and archived upon receipt by CDX, immediately after a successful virus scan; archiving will include a digital signing of the files by EPA to ensure file integrity;
- The electronic document as it was signed with its submitter digital signature affixed; these will be captured after the digital signatures are verified, and will include data generated by the verification process;
- The electronic document as it was signed, with the verified digital signature affixed, the date and time of receipt and EPA's digital signature of the entire content; this will constitute the "copy of record"; and
- The submission acknowledgments sent back to the submitter with EPA signatures, including the data and time these are transmitted.

If, at a later date, there is a question about the file that was received, the EPA can use this sequence of archived files to verify that no changes have been made to the original input from the submitter. Of course, we believe the fact that these archived files are digitally signed will make it impossible for any of these files to be modified without detection. As noted earlier, a digital signature is a function of the "message digest" or "hash" of the document or file it is used to sign. Any modification to the file would change its "hash"--which will be different for each variation of the file--and this would automatically invalidate the signature. A change in even a single character of a file or document would invalidate its digital signature, and would trigger an error warning when processed by the CDX server.

In terms of archive storage, the CDX will archive to multiple formats: hard disk, tape, and optical media. This use of multiple formats is designed to ensure that degradation of one format would not jeopardize EPA's long-term storage capability for submitted data. The CDX archives will be written out to an nline disk system when they are first created. They will be copied to an off-line disk system and also backed up to magnetic tape every day, with full backups to tape on a weekly basis. The schedule for backup to optical media--and the requirements for rapidity of retrieval--have not yet been decided, and EPA welcomes any suggestions in this area. The optical media archiving is intended to provide for long-term storage, extending to periods of 20-50 years.

Finally, CDX will also provide security for data exchanges. To protect lient-server transactions, including the report submission and transmission of acknowledgments, CDX will use a protocol that encrypts the files being exchanged between a "client" PC and the CDX server while these files travel through the network. In addition, the private key, as already noted, will be password protected; it will also provide separate password protection of access to the private key that generates the digital signature. To further protect a user's account from theft or spurious use by an intruder across a company network, current planning calls for the CDX client software to be "localized" to the particular PC on which it is installed--preventing access to this software installed on a particular PC from other PCs connected to it via a network. It is worth adding that, when the private key is created--in connection with the registration process--this can be done in a way that prohibits its export. If this option is invoked, the private key can never be moved--whether to a floppy or to another computer--so if a signature-holder had to move to another machine, the existing public/private key pair assigned to this individual will have to be abandoned, and he or she will have to apply for a new certificate. While EPA is not currently planning to require this option, we are seeking comment both on whether it would involve too much burden for users and on whether the option is necessary to protect the private key from compromise.

3. Electronic Data Interchange (EDI) Standards

As discussed in section IIA, above, EPA has, historically, based its approach to electronic reporting on EDI standards, specifically those developed and maintained under ANSI ASC X12. Today's proposal represents a departure from this approach, in that the regulatory language itself does not specify any particular data formats or transaction set standards. In addition, as already noted, the system that EPA is proposing to use in implementing electronic reporting--the `Central Data Exchange'--will not specify ANSI X12 standards as the only syntax for automated transfers of compliance data. Nonetheless, the

EDI standards on which we have relied in the past will still serve to define many of the data sets that we expect CDX to accept from our submitters.

There are two reasons for this. The first is simply that a significant minority of very large company submitters conduct their electronic commerce using ANSI-based EDI; we want to be able to accommodate these companies and allow them to conduct their transactions with CDX using the same infrastructure they use in commerce. The second reason, is generally that ANSI standards continue to provide a precise, well-documented and widely-recognized way of describing the structure of electronic transactions--including the elements of data involved and how they are related to each other. By providing this clarity, these standards-based descriptions facilitate the implementation of an electronic transfer even where ANSI X12 is replaced by another format for the data files--that is, another way of ordering, grouping, labeling and separating the elements of data. In addition, many of the commercial off-the-shell (COTS) electronic commerce products can translate X12 syntax into other formats, such as "extended mark-up language" (XML).

CDX will make EDI available for many, if not all, of the reports and other documents it is set up to receive. Beyond issues of configuring the CDX server software to recognize and process EDI-formatted files, implementation of EDI is largely a matter of developing the implementation guidance for each of the environmental reports to be supported. As noted in Section II.A of this preamble, the implementation guidance does three things. First, it addresses such procedural matters as: interactions with the communications network (which, under current plans, can be a 'value-added network' or 'VAN', but can also be the Internet), schedule for submissions and acknowledgments, transaction records to be maintained, and so on. Second, it stipulates the specific ANSI X12 standard file transmission formats--that is, "transaction sets"--to be used for the specified reports. Third, the guidance specifies how the stipulated transaction sets being used are to be interpreted as they are applied to the environmental report in question.

As noted in Section II.A, X12 transaction sets are generic in the sense that they typically leave a number of their components as 'optional', and use data-element specifications that are open to multiple interpretations. Therefore the implementation guidance must, at the very least, establish the correlation between the generic data elements and the specific data elements in the EPA report that would be put into this format--in essence, this is to specify which data field in the EPA report goes where in the transaction set format. This is sometimes described as mapping the generic transaction set to the particular set of data elements it will serve to format. The result of this "mapping" process is often referred to as the "implementation convention" (IC) of the transaction set for the report or document in question. Accordingly, each EPA program-specific implementation guidance will include the applicable ICs.

EPA has written and codified ICs for many of the Agency's major compliance reports, and several more are under development. These ICs have been (or will be) approved as a 'Federal Implementation Convention'. This approval process, which involves public notice and comment, is managed by the Federal Electronic Data Interchange Standards Management Coordinating Committee (FESMCC), under the Federal Information Processing Standard Publication (FIPS PUB) 161-2, entitled "lectronic Data Interchange." All approved Federal IC's are registered with the National Institute of Standards and Technology (NIST). The NIST registry, now including 863E, is posted at: http://snad.ncsl.nist.gov/fededi/. Whenever EPA intends to upgrade to a new version or release of the ANSI X12 standards, or in any other way modify the applicable IC, EPA will give notice of its intent in the Federal Register and will establish a conversion date. Affected regulated entities will then have a minimum of sixty (60) calendar days from the conversion date to conform to the modified IC; EPA will discontinue support of the previous version of the IC no sooner than ninety (90) calendar days after the conversion date.

The full list of currently approved ICs is:

- 863E--Report of Test Results (Discharge Monitoring Report): This IC is available in PDF, RTF, ASCII, SEF formats for Version 4010 from http://snad.ncsl.nist.gov/dartg/edi/4010-ic.html.
- The 863S--Report of Test Results (Safe Drinking Water) IC is currently in the FESMCC approval process. When approved, it will be available in PDF, RTF, ASCII, SEF formats for Version 4010.
- In addition, ANSI ASC X12 has recently approved a new transaction set specifically developed by EPA to support environmental reporting, the 179. The 179 consolidates several EPA reports into a single transaction set. The 179 can convey a Discharge Monitoring Report, Hazardous Waste Report, Toxic Release Inventory report, the Air Emission Inventory report, or Risk Management Plan. The 179 was published initially in the ANSI ASC Version 4031. The ICs for the 179 are being developed and will coordinated through the FESMCC process and published on the NIST web site after approval.

4. The Transaction Environment

As explained in earlier sections, CDX would allow submitters to transmit data either through automated file transfer, or via on-screen "smart forms" provided as a part of the downloaded "desktop". In either case, however, the signature/certification "scenario"--that is, the series of steps surrounding the digital signing of the report--will be the same, consisting of: a data review sequence; the signature process; and an acknowledgment sequence.

These steps will largely be governed by operation of the CDX software, and the interaction of the client PC with the CDX server. Taking these in order, data review will take place online, with the CDX server providing the transmitted data for submitter review in a format that is easily read and understood, possibly with a visual layout similar to the applicable paper form (if there is one). The server will present the data one screen at a time--downloaded to the client browser--and it will not allow the submitter to initiate the signing process until the last screen has appeared. The review sequence will end when the submitter clicks a button at the bottom of the last data screen to initiate signature.

Once initiated, the signature process will first display the certification statement, certifying to the truth of the data to be submitted, and also including a warning that by initiating the signing process the submitter agrees that he or she is using the signature in compliance with the signature agreement that was signed when the signature device was issued. The exact content and wording of the first of these statements will be consistent with the language suggested for this purpose in sub-section IV.D.4 of this preamble. In any event, the submitter will be prompted to click agreement with this statement, after which the submitter will be prompted to enter his or her password launching the digital signature process. The digital signature will be created by using the submitter's private key to encrypt a 'hash' of all the elements of the screens the submitter has reviewed--including screen layout, data field labels, data elements, and certification statements. Once the signature is created and affixed, the signed report will be immediately transmitted to the server.

Transmission to the server will initiate the acknowledgment sequence. Upon receipt of the transmission, CDX will automatically create an acknowledgment that includes the date and time of receipt. This acknowledgment will be posted to the submitter's password-protected mailbox on the server, and/or to a submitter-specified email address. In addition, the server will also create a "copy of record" of the submission, by applying an EPA digital signature to the entire file received, including the submitter's digital signature. EPA will count this "copy of record" as the "original" of the submission for all legal purposes, and will maintain this electronic document in the CDX archive. As currently planned, this "copy of record" will be placed in the submitter's password-protected mailbox on the server. When the submitter next logs into CDX, the first screen he or she sees will present the list of copies of record (and acknowledgments, unless these are sent by email) that currently await submitter review; the submitter will be able to download and archive these documents. Of course, the submitter will be encouraged to review these copies of record to confirm that they correspond with what he or she intended to submit, and to notify EPA immediately in the case of any discrepancy.

In our design of this three-part scenario (data review, signature process, and acknowledgment), our major goals have been to make CDX simple, intuitive and easy for submitters to use, while--at the same time--ensuring that a submitter knows and understands what he or she is certifying, the meaning of affixing a digital signature to the electronic document, what has happened, and what EPA considers to be the document that was submitted. EPA seeks comment on the appropriateness of these goals and whether more or less should be designed into CDX to ensure that it meets these goals.

VI. Regulatory Requirements

A. Executive Order 12866

Pursuant to the terms of Executive Order 12866 (58 FR 51735, October 4, 1993), it has been determined that this rule is a "significant regulatory action" because it raises novel legal and /or policy issues. As such, this action was submitted to OMB for review. Changes made in response to OMB suggestions or recommendations will be documented in the public record.

B. Executive Order 13132

Executive Order 13132, entitled "Federalism" (64 FR 43255, August 10, 1999), requires EPA to develop an accountable process to ensure "meaningful and timely input by State and local officials in the development of regulatory policies that have federalism implications." "Policies that have federalism implications" is defined in the Executive Order to include regulations that have "substantial direct effects on the States, on the relationship between the national government and the States, or on the distribution of power and responsibilities among the various levels of government."

Under Section 6 of Executive Order 13132, EPA may not issue a regulation that has federalism implications, that imposes substantial direct compliance costs, and that is not required by statute, unless the Federal government provides the funds necessary to pay the direct compliance costs incurred by State and local governments, or EPA consults with State and local officials early in the process of developing the proposed regulation. EPA also may not issue a regulation that has federalism implications and that preempts State law, unless the Agency consults with State and local officials early in the process of developing the proposed regulation.

This proposed rule does not have federalism implications. It will not have substantial direct effects on the States, on the relationship between the national government and the States, or on the distribution of power and responsibilities among the various levels of government, as specified in Executive Order 13132. The proposed rule would not require States to accept electronic reports. The effect of this rule would be to provide additional regulatory flexibility to States because States could choose to accept electronic data in satisfaction of EPA reporting requirements. Authorized States that did choose to accept electronic reports under this rule would incur expenses initially in developing systems or modifying existing systems to meet the criteria in this rule. However, the Cost/Benefit analysis associated with this proposed rule, summarized in section IV.E of this preamble, estimates that States' overall cost savings from implementing electronic reporting will more than compensate for these initial expenses. Additionally, EPA believes that even in the absence of this proposed rule, States' implementing electronic reporting on their own initiative would generally choose to meet the criteria that this rule proposes. Thus, the requirements of section 6 of the Executive Order do not apply to this rule. Although section 6 of Executive Order 13132 does not apply to this rule, EPA did consult with State and local officials in developing this rule.

C. Paperwork Reduction Act

The information collection requirements in this proposed rule have been submitted for approval to the Office of Management and Budget (OMB) under the Paperwork Reduction Act (PRA), 44 U.S.C. 3501 et seq. An Information Collection Request (ICR) document has been prepared by EPA (ICR No. 2002.02) and a copy may be obtained from Sandy Farmer by mail at Collection Strategies Division; U.S. Environmental Protection Agency (2822); 1200 Pennsylvania Ave., NW, Washington, DC 20460, by email at farmer.sandy@epamail.epa.gov, or by calling (202) 260-2740. A copy may also be downloaded off the Internet at http://www.epa.gov/icr.

The proposed rule would allow reporting entities to voluntarily submit reports and other information electronically, thereby streamlining and expediting the process for reporting. It will also allow facilities to maintain electronic records for information/data currently required by regulation or statute to be maintained by the regulated entity onsite. EPA is proposing this rule on cross-media electronic reporting and record-keeping, in part, under the authority of the Government Paperwork Elimination Act, Public Law 105-277, which amends the PRA.

The CROMERRR ICR primarily covers the registration information which will be collected from individuals wishing to submit electronic reports on behalf of a regulated entity and will be used to establish the identity of that individual and the regulated entity he or she will represent. It also covers activities incidental to electronic reporting. Submission of reports in an electronic format will be voluntary.

The total annual reporting and record-keeping burden this ICR estimates for all facilities is 874,853 hours, which includes the tasks of collecting data, managing the system, and keeping records. A more detailed description of these activities includes the following: registering with EPA or State electronic document receiving systems, including invitation, verification, certificate issuance, and access and agreement; renewing registration with the electronic document receiving system once every two years; activities related to maintaining the electronic signature, including renewing the signature certificate, reporting loss, theft, or other compromise of any component of an electronic signature, and surrender of electronic signature; and facility electronic record-keeping, including generating and maintaining complete e-records and documents. It is expected that tasks associated with system registration will take an average of one (1) hour per registrant/entity and the estimated number of likely respondents is 324,370. For the first year, there will be start-up and annual operation and maintenance (O&M) costs. Costs for the following two years will only involve annual O&M, based on the assumption that the registration will be valid for three years. Total annual start-up costs are estimated at $10,700,000.00 and annual O&M costs are estimated at $5,100,123.96.

Burden means the total time, effort, or financial resources expended by persons to generate, maintain, retain, or disclose or provide information to or for a Federal agency. This includes the time needed to review instructions; develop, acquire, install, and utilize technology and systems for the purposes of collecting, validating, and verifying information, processing and maintaining information, and disclosing and providing information; adjust the existing ways to comply with any previously applicable instructions and requirements; train personnel to be able to respond to a collection of information; search data sources; complete and review the collection of information; and transmit or otherwise disclose the information. An Agency may not conduct or sponsor, and a person is not required to respond to a collection of information unless it displays a currently valid OMB control number. The OMB control numbers for EPA's regulations are listed in 40 CFR part 9 and 48 CFR chapter 15.

Comments are requested on the Agency's need for this information, the accuracy of the provided burden estimates, and any suggested methods for minimizing respondent burden, including through the use of automated collection techniques. Send comments on the ICR to the Director, Collection Strategies Division; U.S. Environmental Protection Agency (2822); 1200 Pennsylvania Ave., NW., Washington, DC 20460; and to the Office of Information and Regulatory Affairs, Office of Management and Budget, 725 17th St., NW., Washington, DC 20503, marked ``Attention: Desk Officer for

EPA." Include the ICR number in any correspondence. Since OMB is required to make a decision concerning the ICR between 30 and 60 days after August 31, 2001, a comment to OMB is best assured of having its full effect if OMB receives it by October 1, 2001. The final rule will respond to any OMB or public comments on the information collection requirements contained in this proposal.

D. Regulatory Flexibility Act

The Regulatory Flexibility Act (RFA), 5 U.S.C. 601 et seq., provides that, whenever an agency promulgates a proposed rule under section 553 of the Administrative Procedures Act, after being required by that section or any other law to publish a general notice of rulemaking, the agency generally must prepare an initial regulatory flexibility analysis (IRFA). The agency must prepare a Final Regulatory Flexibility Analysis (FRFA) for a final rule unless the head of the agency certifies that it will not have a significant economic impact on a substantial number of small entities.

Today's rule is not subject to the RFA because electronic reporting and record-keeping is voluntary and will only apply to those States and tribes that seek EPA approval to allow electronic reporting and record-keeping under their authorized programs and to regulated entities that seek to maintain records or transmit compliance reports electronically to EPA or authorized/approved States or tribes. These changes will reduce the burden on all affected entities, including small businesses. Accordingly, this rule is certified as having no Significant economic impact on a substantial number of small businesses. Respondent burden is the burden placed upon each individual reporting entity involved in set up, configuration and implementation of electronic submission of environmental compliance reports. Regulated entities will find that the initial set up process requires some expenditure of time and resources, but in the long run, this process will reduce the time spent on submissions each year. The Cost/Benefit analysis associated with this proposed rule, summarized in section IV.E, estimates that electronic reporting and record-keeping, when fully implemented, will reduce regulated facility compliance cost by more than $300 million per year. The Administrator therefore certifies, pursuant to section 605(b) of the RFA, that this rule will not have a significant economic impact on a substantial number of small entities.

E. Unfunded Mandates Reform Act

Title II of the Unfunded Mandates Reform Act of 1995 (UMRA), Public Law 104-4, establishes requirements for Federal agencies to assess the effects of their regulatory actions on State, local, and tribal governments and the private sector. Under section 202 of the UMRA, EPA generally must prepare a written statement, including a cost-benefit analysis, for proposed and final rules with "Federal mandates" that may result in expenditures to State, local, and tribal governments, in the aggregate, or to the private sector, of $100 million or more in any one year. Before promulgating an EPA rule for which a written statement is needed, section 205 of the UMRA generally requires EPA to identify and consider a reasonable number of regulatory alternatives and adopt the least costly, most cost-effective or least burdensome alternative that achieves the objectives of the rule. The provisions of section 205 do not apply when they are inconsistent with applicable law. Moreover, section 205 allows EPA to adopt an alternative other than the least costly, most cost-effective or least burdensome alternative if the Administrator publishes with the final rule an explanation why that alternative was not adopted.

Before EPA establishes any regulatory requirements that may significantly or uniquely affect small governments, including tribal governments, it must have developed under section 203 of the UMRA a small-government agency plan. The plan must provide for notifying potentially affected small governments, enabling officials of affected small governments to have meaningful and timely input in the development of EPA regulatory proposals with significant Federal intergovernmental mandates, and informing, educating, and advising small governments on compliance with the regulatory requirements.

The Agency has determined that this rule does not contain a Federal mandate that may result in expenditures of $100 million or more for State, local and tribal governments, in the aggregate, or the private sector in any one year. Today's rule provides additional flexibility to the States in complying with current regulatory requirements and reduces the burden on affected governments. Thus, today's rule is not subject to the requirements in sections 202 and 205 of the UMRA.

The Agency has determined that this rule contains no regulatory requirements that might significantly or uniquely affect small governments and thus this rule is not subject to the requirements in section 203 of UMRA. This rule will not significantly affect small governments because it provides additional flexibility in complying with pre-existing regulatory requirements.

F. National Technology Transfer and Advancement Act

Section 12(d) of the National Technology Transfer and Advancement Act of 1995 (``NTTAA''), Public Law 104-113, section 12(d) (15 U.S.C. 272 note) directs EPA to use voluntary consensus standards in its regulatory activities unless to do so would be inconsistent with applicable law or otherwise impractical. Voluntary consensus standards are technical standards (e.g., materials specifications, test methods, sampling procedures, and business practices) that are developed or adopted by voluntary consensus standards bodies. The NTTAA directs EPA to provide Congress, through OMB, explanations when the Agency decides not to use available and applicable voluntary consensus standards.

This rulemaking involves information technology standards for electronic formats and for electronic signatures. EPA is exploring a number of standards-based approaches to Web forms, including electronic data exchange formats based upon the American National Standards Institute (ANSI) Accredited Standards Committee's (ASC) X12 for Electronic Data Interchange or EDI. EPA is also proposing Internet data exchange formats based on the Extensible Mark-up Language (XML) specifications developed by the World Wide Web Consortium (W3C). The World Wide Web Consortium, however, is not a voluntary consensus standards body within the meaning of the NTTAA, and EPA could not identify an applicable consensus standard for creating and transmitting data using XML. Therefore, EPA has decided to propose an XML data exchange format, referred to as a document type definition for Internet transmissions as an alternative to the ANSI ASC X12 formats that are customarily transmitted across Value Added Networks. It is possible that the ANSI ASC X12 standards body will develop standards for XML document definitions in the future, and EPA will monitor this situation as we develop a final rulemaking.

G. Executive Order 13045

The Executive order, Protection of Children from Environmental Health Risks and Safety Risks (62 FR 19885, April 23, 1997) applies to any rule that EPA determines (1) "economically significant" as defined under Executive Order 12866 and (2) concerns an environmental health or safety risk that EPA has reason to believe may have a disproportionate effect on children. EPA interprets the Executive Order 13045 as encompassing only those regulatory actions that are risk-based or health-based, such that the analysis required under section 5-501 of the Executive Order has the potential to influence the regulation.

This rule is not subject to Executive Order 13045 because it is not an economically significant action as defined by Executive Order 12866 and it does not involve decisions regarding environmental health or safety risks. This rule develops technical procedures for the voluntary submission of environmental compliance data electronically.

H. Executive Order 13175

Executive Order 13175, entitled, "A Consultation and Coordination with Indian Tribal Governments" (65 FR 67249, November 6, 2000), requires EPA to develop an accountable process to ensure "meaningful and timely input by tribal officials in the development of regulatory policies that have tribal implications." "Policies that have tribal implications" is defined in the Executive Order to include regulations that have "substantial direct effects on one or more Indian tribes, on the relationship between the Federal government and the Indian tribes, or on the distribution of power and responsibilities between the Federal government and Indian tribes."

This proposed rule does not have tribal implications. It will not have substantial direct effects on tribal governments, on the relationship between the Federal government and Indian tribes, or on the distribution of power and responsibilities between the Federal government and Indian tribes, as specified in Executive Order 13175. The proposed rule would not require Indian tribes to accept electronic reports. The effect of this rule would be to provide additional regulatory flexibility to Indian tribes because tribes could choose to accept electronic data in satisfaction of EPA reporting requirements. Authorized tribal programs that did choose to accept electronic reports under this rule would incur expenses initially in developing systems or modifying existing systems to meet the criteria in this rule. However, the Cost/Benefit analysis associated with this proposed rule, summarized in section IV.E of this preamble, estimates that tribes' overall cost savings from implementing electronic reporting will more than compensate for these initial expenses. Additionally, EPA believes that even in the absence of this proposed rule, Indian tribes' implementing electronic reporting on their own initiative would generally choose to meet the criteria that this rule proposes. Thus, Executive Order 13175 does not apply to this rule. In the spirit of Executive Order 13175, and consistent with EPA policy to promote communications between EPA and tribal governments, EPA specifically solicits additional comment on this proposed rule from tribal officials.

I. Executive Order 13211 (Energy Effects)

This rule is not a "significant energy action" as defined in Executive Order 13211, "Actions Concerning Regulations That Significantly Affect Energy Supply, Distribution, or Use" (66 FR 28355 (May 22, 2001)) because it is not likely to have a significant adverse effect on the supply, distribution, or use of energy. EPA has concluded that this rule is not likely to have any adverse energy effects.

The Proposed Rule

Therefore, it is proposed that title 40 chapter I of the Code of Federal Regulations be amended by adding a new part 3, and revising parts 51, 60, 63, 70, 123, 142, 145, 162, 233, 257, 258, 271, 281, 403, 501, 745, and 763 to read as follows:

PART 3--[NEW]
ELECTRONIC REPORTING; ELECTRONIC RECORDS

Subpart A--General Provisions
Sec.
3.1 Scope.
3.2 Implementation.
3.3 Definitions.
3.4 [Reserved]
Subpart B--Electronic Reporting to EPA
3.10 What are the requirements for acceptable electronic documents?
3.20 How will EPA provide notice of changes to the Central Data Exchange?
3.30 [Reserved]
Subpart C--Electronic Record-keeping Under EPA Programs
3.100 What are the requirements for acceptable electronic records?
3.200 [Reserved]
Subpart D--Electronic Reporting and Record-keeping Under EPA-Approved State Programs
3.1000 How are authorized State, tribal or local environmental programs modified to allow electronic reporting?
3.2000 What are the criteria for acceptable electronic document receiving systems?
3.3000 How are authorized State, tribal or local environmental programs modified to allow electronic record-keeping?
3.4000 [Reserved]

Authority: 7 U.S.C. 136 to 136y; 15 U.S.C. 2601 to 2692; 33 U.S.C. 1251 to 1387; 33 U.S.C. 1401 to 1445; 33 U.S.C. 2701 to 2761; 42 U.S.C. 300f to 300j-26; 42 U.S.C. 6901-6992k; 42 U.S.C. 7401 to 7671q; 42 U.S.C. 9601 to 9675; 42 U.S.C. 11001 to 11050; 15 U.S.C. 7001; 44 U.S.C. 3504 to 3506.

Subpart A--General Provisions

Sec. 3.1 Scope.

What Is Covered by This Part?

(a) This part sets forth the conditions under which EPA will accept the submission of electronic reports and other electronic documents, as well as the maintenance of electronic records, by regulated entities, as satisfying requirements under this Title to submit reports or other documents, or to keep records. This part also sets forth the standards and process for EPA approval of changes to authorized State, tribal, and local environmental programs to allow electronic report or document submission or electronic record maintenance in satisfaction of requirements under such authorized programs. This part does not require submission of electronic reports or documents or electronic recordkeeping in lieu of paper. This part confers no right or privilege to submit or maintain data electronically and does not obligate EPA, or State, tribal or local agencies to accept electronic data.

(b) Subpart C of this part applies to records in electronic form that are created, modified, maintained, archived, retrieved, or transmitted by regulated entities under any recordkeeping requirements under this Title. However, Subpart C of this part does not provide for the conversion of existing paper documents or records into electronic form. Subpart C of this part also does not apply to the Agency's recordkeeping requirements set forth in regulations governing contracts, grants, and financial management programs.

Sec. 3.2 Implementation.

What Requirements May Be Satisfied by Electronic Reporting and Electronic Recordkeeping?

(a) Electronic reporting to EPA. Any requirement in this Title that a document be created and transmitted or otherwise provided to EPA may be satisfied with an electronic document, in lieu of a paper document, provided that:
　(1) The electronic document satisfies the requirements of Sec. 3.10; and
　(2) EPA has published a notice in the Federal Register announcing that EPA is prepared to receive in electronic form documents required or permitted by the named Part or Subpart of this Title.

(b) Electronic recordkeeping under EPA programs. Except as provided under paragraph (d) of this section or excluded under Sec. 3.1(b), any requirement in this Title that a record be maintained may be satisfied by maintaining an electronic record, in lieu of a paper record provided that:
　(1) The electronic record satisfies the requirements of Sec. 3.100; and

(2) EPA has published a notice in the Federal Register announcing that EPA is prepared to recognize electronic records under the named Part or Subpart of this Title.

(c) Electronic reporting and recordkeeping under an EPA-authorized State, tribal, or local environmental program. Except as provided under paragraph (d) of this section, any requirement under authorized State, tribal, or local environmental programs that reports or documents be submitted or records be maintained may be satisfied with electronic report or document submission, or with electronic record maintenance, respectively, provided that: EPA has approved, in accordance with Subpart D of this part, the changes to the authorized State, tribal, or local environmental program to allow the electronic report or document submission or the electronic record maintenance in satisfaction of the authorized program requirement.

(d) Limitation on the use of electronic records under EPA programs and EPA-authorized State, tribal, or local environmental programs. Electronic records that meet the requirements of this Part may be used in lieu of paper records unless paper records are specifically required by other provisions in this Title that take effect on or after [date of promulgation of this regulation].

Sec. 3.3 Definitions.

What definitions are applicable to this part? The definitions set forth in this section apply when used in this part.

- Acknowledgment means a confirmation of document receipt.
- Administrator means the Administrator of the Environmental Protection Agency.
- Agency means the Environmental Protection Agency or a State, tribal, local or other federal agency that administers a federal environmental program under this Title.
- Agency electronic signature means an electronic signature of an individual who is authorized to sign an electronic document on an agency's behalf.
- Authorized State, Tribal, or local environmental program means an environmental program which EPA has approved, authorized, or delegated to a State, tribe or local government to administer under a federal environmental program.
- Communicate means to successfully and accurately convey a document, data, or information from one entity to another.
- Electronic document means a document that is submitted to an agency or third-party as an electronic record, and communicated via a telecommunications network. For purposes of this part, electronic

document excludes documents submitted on such magnetic media as diskettes, compact disks or tapes; it also excludes facsimiles.
- Electronic document receiving system means any set of apparatus, procedures, software, records or documentation used to receive documents communicated to it via a telecommunications network.
- Electronic record means any combination of text, graphics, data, audio, pictorial, or other information represented in digital form that is created, modified, maintained, archived, retrieved or distributed by a computer system.
- Electronic record-retention system means any set of apparatus, procedures, software, records or documentation used to retain exact electronic copies of electronic records and electronic documents.
- Electronic submission mechanism means any set of apparatus, procedures, software, records or documentation used to communicate an electronic document to an electronic document receiving system.
- Electronic signature means any electronic record that is incorporated into (or appended to) an electronic document for the purpose of expressing the same meaning and intention that an individual's handwritten signature would express if affixed in the same relation to the document's content presented on paper.
- Electronic signature device means a code or other mechanism that is used to create electronic signatures. Where the device is used to create an individual's electronic signature, then the code or mechanism must uniquely belong to or be associated with or assigned to that individual. Where the device is used to create an organization's electronic signature, then the code or mechanism must uniquely belong to or be associated with or assigned to that organization.
- EPA means the United States Environmental Protection Agency.
- Handwritten signature means the scripted name or legal mark of an individual, handwritten by that individual with a writing or marking instrument such as a pen or stylus and executed or adopted with the present intention to authenticate a writing in a permanent form. The physical instance of the scripted name or mark so created constitutes the handwritten signature. The scripted name or legal mark, while conventionally applied to paper, may also be applied to other hard media.
- Metadata means data that describes the properties of other data or collections of data (e.g., a database); with respect to a database or file containing data, metadata could include information about the database's structure, the date and time that data was created or added or changed, definitions of the data elements, descriptions of the accuracy of the data, etc.

- Receive means to successfully acquire electronic documents in a format that can be processed by the receiving system.
- Regulated entity means any entity that maintains records or submits documents to EPA to satisfy requirements under this Title, or that maintains records or submits documents to a State, tribal, or local agency to satisfy requirements under programs authorized under this Title. A State, tribal, or local agency or tribe may be a regulated entity where it maintains records or submits documents to satisfy requirements that apply to it under this Title (including regulations governing authorized State, tribal, or local programs); a State, tribal, or local agency will not be a regulated entity where it maintains records or submits documents exclusively for other purposes, for example as a part of administrative arrangements between States and EPA to share data.
- Submit means to communicate a document so that it is received by the intended recipient.
- Third-party system means an electronic document receiving system that is owned or operated by an entity that is neither a submitter of the electronic documents the system receives nor an agency to which these electronic documents are submitted.

Sec. 3.4 [Reserved]

Subpart B--Electronic Reporting to EPA

Sec. 3.10 What are the requirements for acceptable electronic documents?

(a) An electronic document will satisfy a federal environmental reporting requirement or otherwise substitute for a paper submission permitted or required under this Title only if:

(1) The electronic document is submitted to an electronic document receiving system as provided under paragraph (b) of this section, and

(2) The electronic document bears valid electronic signatures, as provided in paragraphs (c), (d) and (e) of this section, to the same extent that the paper submission for which it substitutes would bear handwritten signatures.

(b) Electronic documents submitted to EPA to satisfy a federal environmental reporting requirement or otherwise substitute for a paper submission permitted or required by a federal environmental program must be submitted to either:

(1) EPA's Central Data Exchange; or

(2) Another EPA electronic document receiving system that the Administrator may designate for the receipt of specified submissions.

(c) An electronic signature is valid if and only if:

(1) The electronic signature is created by a person who is authorized to sign the document, with an electronic signature device that this person is authorized to use; and

(2) The electronic signature meets the validation requirements of the electronic document receiving system to which it is submitted.

(d) A valid electronic signature on any electronic document submitted to atisfy a federal or federally authorized State, tribal or local government environmental reporting requirement legally binds or obligates the signatory, or makes the signatory responsible, to the same extent as the signatory's hand-written signature on a paper document submitted to satisfy the same federal or federally authorized environmental reporting requirement.

(e) Proof that an individual's electronic signature was affixed to an electronic document is evidence, and may suffice to establish, that the individual who was issued that signature affixed the signature and did so with the intent to sign the electronic document to give it effect.

Sec. 3.20 How will EPA provide notice of changes to the Central Data Exchange?

(a) Except as provided under paragraph (b) of this section, whenever EPA plans to change Central Data Exchange hardware or software in ways that would affect the submission process:

(1) Where the equipment, software or services needed to submit electronic reports to the Central Data Exchange would be changed, EPA will provide public notice and seek comment on the proposed change at least a year in advance of the proposed implementation data;

(2) Otherwise, EPA will provide public notice at least sixty (60) days in advance of implementation.

(b) Any change which the Administrator determines is needed to ensure the security and integrity of the Central Data Exchange is exempt from the provisions of paragraph (a) of this section. However, to the extent consistent with ensuring the security and integrity of the system, EPA will provide public notice of any change to the Central Data Exchange made under the authority expressly reserved by this subsection.

Sec. 3.30 [Reserved]

Subpart C--Electronic Recordkeeping under EPA Programs

Sec. 3.100 What are the requirements for acceptable electronic records?

(a) An electronic record or electronic document will satisfy a recordkeeping requirement of an EPA-administered federal environmental program under this Title only if it is generated and maintained by an acceptable electronic record-retention system as specified under this subsection. For purposes of maintaining electronic records that satisfy recordkeeping requirements under this Title, an acceptable electronic record-retention system must:

(1) Generate and maintain accurate and complete electronic records and electronic documents in a form that may not be altered without detection;

(2) Maintain all electronic records and electronic documents without alteration for the entirety of the required period of record retention;

(3) Produce accurate and complete copies of any electronic record or electronic document and render these copies readily available, in both human readable and electronic form, for on-site inspection and off-site review, for the entirety of the required period of record retention;

(4) Provide that any electronic record or electronic document bearing an electronic signature contain the name of the signatory, the date and time of signature, and any information that explains the meaning of the affixed signature;

(5) Prevent an electronic signature that has been affixed to an electronic record or electronic document from being detached, copied, or otherwise compromised;

(6) Use secure, computer-generated, time-stamped audit trails that automatically record the date and time of operator entries and actions that create, modify, or delete electronic records or documents;

(7) Ensure that record changes do not obscure previously recorded information and that audit trail documentation is retained for a period at least as long as that required for the subject electronic records or electronic documents to be available for agency review;

(8) Ensure that electronic records and electronic documents are searchable and retrievable for reference and secondary uses, including inspections, audits, legal proceedings, third party disclosures, as required by applicable regulations, for the entirety of the required period of record retention; and

(9) Archive electronic records and documents in an electronic form which preserves the context, meta data, and audit trail, and, if required, must ensure that:

(i) Complete records can be transferred to a new system;

(ii) Related meta data can be transferred to a new system; and

(iii) Functionality necessary for use of records can be reproduced in new system;

(b) Computer systems (including hardware and software), controls, and attendant documentation maintained under this Part must be readily available for, and subject to, agency inspection.

(c) Where electronic records bear electronic signatures that meet the requirements in paragraphs (a)(4) and (a)(5) of this section, EPA will consider the electronic signatures to be equivalent to full handwritten signatures, initials, and other general signings as required by federal or federally authorized State, tribal or local government environmental regulations, unless specifically excepted by regulations(s) effective on or after [date of promulgation of this regulation].

Sec. 3.200 [Reserved]

Subpart D--Electronic Reporting and Recordkeeping Under EPA-Approved State Programs

Sec. 3.1000 How are authorized State, tribal or local environmental programs modified to allow electronic reporting?

(a) State, tribes, or local environmental programs that wish to receive electronic reports or documents in satisfaction of requirements under such programs must revise or modify the EPA-approved State, tribal, or local environmental program to ensure that it meets the requirements of this part. The State, tribe, or local government must use existing State, tribal, or local environmental program procedures in making these program revisions or modifications.

(b) In order for EPA to approve a program revision under paragraph (a) of this section the State, tribe, or local government must demonstrate that electronic reporting under this program will:

(1) Use an acceptable electronic document receiving system as specified under Sec. 3.2000;

(2) Require that any electronic report or document must bear valid electronic signatures, as provided in Sec. 3.10(c), (d) and (e), to the same extent that the paper submission for which it substitutes would bear handwritten signatures under the State, tribal, or local environmental program.

Sec. 3.2000 What are the criteria for acceptable electronic document receiving systems?

An electronic document receiving system that is acceptable for purposes of receiving electronic reports or documents submitted under provisions of an authorized State, tribal or local environmental program must meet all of the following requirements:

(a) General system-security. An acceptable electronic document receiving system must:

(1) Have strong and effective protections against unauthorized access to the system;

(2) Have strong and effective protections against the unauthorized use of any electronic signature on electronic documents submitted or received;

(3) Provide for the detection of unauthorized access or attempted access to the system and unauthorized use or attempted use of any electronic signature on electronic documents submitted or received;

(4) Prevent the modification of an electronic document once an electronic signature has been affixed;

(5) Ensure that the electronic documents and other files necessary to meet the requirements under paragraphs (f) and (g) of this section are protected from modification or deletion;

(6) Ensure that the system clock is accurate and protected from tampering or other compromise; and

(7) Have strong and effective protections against any other foreseeable corruption or compromise of the system.

(b) Validity of data. An acceptable electronic document receiving system must generate data sufficient to prove, in private litigation, civil enforcement proceedings, and criminal proceedings, that:

(1) The electronic document was not altered in transmission or at any time after receipt;

(2) The electronic document was submitted knowingly and not by accident; and

(3) In the case of documents requiring the signature of an individual, that the document was actually submitted by the authorized signature holder and not some other person.

(c) Electronic signature method. By virtue of its presence as a part of an electronic document submitted or received, an electronic signature must uniquely identify the particular individual who has used it to sign an electronic document or otherwise certify to the truth or accuracy of the document contents; therefore, an acceptable electronic document receiving system must only validate electronic signatures created with a method that:

(1) Meets the registration requirements of paragraph (d) of this section;

(2) Meets the signature/certification requirements of paragraph (e) of this section;

(3) Prevents an electronic signature from being excised, modified, or copied for re-use without detection once it has been affixed to an electronic document by the authorized individual;

(4) Provides protection against the use of a specific electronic signature by unauthorized individuals; and

(5) Ensures that it is impossible to modify an electronic document without detection once the electronic signature has been affixed.

(d) Submitter registration process. An acceptable electronic document receiving system must require that anyone who submits an electronic document to the system first register with the agency to which the document is to be submitted. The registration process must establish the identities of both the registrant, who is the prospective submitter, and any entity that the registrant is authorized to represent, and must establish that the registrant is authorized to submit the document in question for the entity being represented. In addition, where the documents to be received will require signature, the registration process must:

(1) Establish the registrant's identity, and the registrant's relation to any entity for which the registrant will submit electronic documents, with evidence that can be verified by information sources that are independent of the registrant and the entity or entities in question and that would be sufficient to identify the registrant as the signature holder for purposes of supporting litigation consistent with paragraph (b) of this section;

(2) Establish and document a unique correlation between the registrant and the code or device that will constitute or create the electronic signature of the registrant as a submitter;

(3) Require that the registrant sign on paper, or in such other manner or medium as the Administrator in his or her discretion may determine as appropriate for a category of electronic reports, an electronic signature agreement specifying at a minimum that the registrant agrees to:

(i) Protect the electronic signature from unauthorized use, and follow any procedures specified by the agency for this purpose;

(ii) Be held as legally bound, obligated, or responsible by use of the assigned electronic signature as by hand-written signature;

(iii) Where the signature method is based on a secret code or key, maintain the confidentiality of each component of the electronic signature;

(iv) In any case, never to delegate the use of the electronic signature, or in any other way intentionally provide access to its use, to any other individual for any reason; and

(v) Report to the entity specified in the electronic signature agreement, within twenty-four hours of discovery, any evidence of the loss, theft, or other compromise of any component of an electronic signature;

(4) Provide for the automatic and immediate revocation of an electronic signature in the event of:

(i) Any actual or apparent violation of the electronic signature agreement;

(ii) Any evidence that the signature has been compromised, whether or not this is reported by the registrant to whom the signature was issued; or

(iii) Notification from an entity that the registrant is no longer authorized by the entity to submit electronic documents on its behalf;

(5) Require that the registrant periodically renew his or her electronic signature agreement, under terms that the Administrator determines provide adequate assurance that the criteria of paragraphs (a) and (b) of this section are met, taking into account both applicable contractual provisions and industry standards for renewal or re-issuance of signature codes or devices.

(e) *Electronic signature/certification scenario.* An acceptable electronic document receiving system that may be used to accept electronic documents bearing an electronic signature must:

(1) Not allow an electronic signature to be affixed to the electronic document until:

(i) The signatory has been provided an opportunity to review all of the data to be transmitted in an on-screen visual format that clearly associates the descriptions or labeling of the information being requested with the signatory's response and which format is identical or nearly identical to the visual format in which a corresponding paper document would be submitted; and

(ii) A certification statement that is identical to that which would be required for a paper submission of the document appears on-screen in an easily-read format immediately above a prompt to affix the certifying signature, together with a prominently displayed warning that by affixing the signature the signatory is agreeing that he or she is the authorized signature holder--referred to by name--has protected the security of the signature as required by the electronic signature agreement signed under paragraph (d)(3) of this section and is otherwise using the signature in compliance with the electronic signature agreement;

(2) Automatically respond to the receipt of an electronic document with transmission of an electronic acknowledgment that:

(i) States that the signed electronic document has been received, clearly identifies the electronic document received, indicates how the signatory may view and download a copy of the electronic document received from a read-only source, and states the date and time of receipt; and

(ii) Is sent to an address whose access is controlled by password, codes or other mechanisms that are different than the controls used to gain access to the system used to sign/certify and send the electronic document;

(3) Automatically creates an electronic ``copy of record" of the submitted report that includes all the warnings, instructions and certification statements presented to the signatory during the signature/certification scenario as described under paragraph (e)(1) of this section, and that:

(i) Can be viewed by the signatory, in its entirety, on-screen in a human-readable format that clearly and accurately associates all of the information provided by the signatory with the descriptions or labeling of the information that was requested;

(ii) Includes the date and time of receipt stated in the electronic acknowledgment required by paragraph (e)(2) of this section;

(iii) Has an agency electronic signature affixed that satisfies the requirements for electronic signature method under paragraphs (c)(3), (c)(4), and (c)(5) of this section;

(iv) Is archived by the system in compliance with requirements paragraph (g) of this section;

(v) Is made available to the submitter for viewing and down-loading; and

(vi) Is protected from a unauthorized access.

(f) Transaction Record. An acceptable electronic document receiving system must create a transaction record for each received electronic document that includes:

(1) The precise routing of the electronic report from the submitter's computer to the electronic document receiving system;

(2) The precise date and time (based on the system clock) of:

(i) Initial receipt of the electronic document;

(ii) Sending of electronic acknowledgment under paragraph (e)(2) of this section;

(iii) Copy of record created under paragraph (e)(3) of this section;

(3) Copy of record as specified under paragraph (e)(3) of this section.

(g) System archives. An acceptable electronic document receiving system must:

(1) Maintain:

(i) The transaction records specified under paragraph (f) of this section, and

(ii) Records of the system on-screen interface displayed to a user under paragraph (e) of this section that can be correlated to the submission of any particular report (including instructions, prompts, warnings, data formats and labels, as well as the sequencing and functioning of these elements);

(2) Maintain the records specified under paragraph (g)(1) of this section for at least the same length of time as would be required for a paper document that corresponds to the received electronic document, and in a way that:

(i) Can be demonstrated to have preserved them in their entirety without alteration since the time of their creation; and

(ii) Provides access to these records in a timely manner that meets the needs of their authorized users.

Sec. 3.3000 How are authorized State, tribal or local environmental programs modified to allow electronic recordkeeping?

(a) State, tribes, or local environmental programs that wish to allow the maintenance of electronic records or documents in satisfaction of requirements under such programs must revise or modify the EPA-approved State, tribal, or local environmental program to ensure that it meets the requirements of this part. The State, tribe, or local government must use existing State, tribal or local environmental program procedures in making these program revisions or modifications.

(b) In order for EPA to approve a program revision under paragraph (a) of this section the State, tribe, or local government must demonstrate that records maintained electronically under this program will satisfy the requirements under Sec. 3.100 of this part.

Sec. 3.4000 [Reserved]

Editor's Note: Parts of this Federal Register Notice having to do with to whom comments should be addressed and when, Agency contact information, List of subjects, and the notes regarding other Parts of FIFRA affected by the Proposed Rule have been deleted from this Addendum due to length..

INDEXES

Author Index

Albert, Richard P., 28
Beidler, W. T., 152
Bennett, Rodney M., 253
Bouvé, Kathryn S., 188
Brady, S. Scott, 40
Bray, L. D., 152
Brown, John C., 98
Casey, William N., 133
Cypher, Robert L., 241
Dabbs, Dudley, 18
Daniel, Renée J., 44
Dobson, Steven C., 209
Gilding, Thomas J., 139
Goeke, Jane E., 5
Harned, William H., 66
Harris, Thomas C., 164
Hoag, Robert, 34
Huffer, M. Evi, 229
Hummel, Susan V., 198
Keatley, Kendy L., 1, 86
Koopmann, Charles H., 159
Krogh, Carmen, 216

Lehr, Mark J., 241
Liem, Francisca E., 241
Ludwig, Kenneth A., 34
Lynn, Mary E., 79
Malak, Sami, 174
Manalo, Joseph, 98
McCall, Deborah, 174
McDevitt, Edward J., 124
McKilligin, Sharon M., 117
Ollinger, Janet, 220
Paczek, Ted, 18
Palmer, W. H., 13
Reibach, Paul H., 220
Schneiders, Gail E., 98
Speaker, Ann M., 117
Stachoviak, Timothy J., 75
Swidersky, Scott, 220
Thompson, Fate, 54
Thompson, Ron, 18
Walla, Robert D., 109
Wells, Carla, 54
Willard, Tommy, 54

Subject Index

A

Access America, electronic government business, 231
Access Certificates for Electronic Services, certificate issuance, 317-319
Access control in Chromeleon™ system, 71-74
Accessibility management in electronic data archiving, 128
Acknowledgement sequence, 326-327
ACPA and EPA joint projects, 141-142, 195, 201
  See also American Crop Protection Association
"Actions Concerning Regulations That Significantly Affect Energy Supply, Distribution or Use" Executive Order 13211, 333
Acute exposure assessment in FQPA, 152-157
Additional controls, proposed electronic reporting and recordkeeping rule, EPA, 289
Adobe Acrobat in PDF document creation, 135-137, 191-194, 196
Advantage™ record keeping system
  electronic Field Trial Notebook development, 20-26
  generic menu choices, 52-53
  project management assistance, 56-65
Advantages and disadvantages, electronic data capture methods, 36-37
Agency-wide strategy for electronic reporting and recordkeeping, EPA, 232-234
Aggregate risk assessment in FQPA, 152-157
AgrEvo Research Center, use of FieldNotes™, 41-43
Agrochemical database in Astrix agrochemical e-compliance architecture, 31-32
Agrochemical industry use of electronic data capture methods, 19-20
Agrochemical registration process, characteristics, 29-30f
American Crop Protection Association
  electronic data submission for pesticide registration, 139-150
  Joint Data Transfer Steering Group, 211
  pesticide registration process improvement, 144-146
  support for international harmonization, pesticide registration processes, 149
  See also ACPA
Analytical chemistry data, audits, 79-85
Analytical data submission on punch cards, Sandoz Crop Protection, 41
Analytical results reporting in Advantage™, 59
ANSI X12 standards in electronic data interchange, Central Data Exchange, 323-325
Antimicrobials Division, EPA, 175
Application administrator, test plan preparation, 77-78

Archive storage, multiple formats, Central Data Exchange, 323
Archivist of the United States, statement on archiving electronic records, 125
Assessment and development for software, overview, 110-111*f*
Astrix agrochemical e-compliance system, 30-33
Astrix FieldNotes™
  used by AgrEvo Research Center, 41-43
  used by Bayer Corporation, 37-39
Audit trail in electronic record systems, FDA requirement, 91, 246-247
Audit trail log, 259
Audits
  electronic field data, 44-53
  electronically captured analytical chemistry data, 79-85
  equipment and systems, 107
  guidance documents for electronically captured data, 80
Authentication in electronic transmission, legal status, 234
Authorized State and tribal programs, provision for electronic reporting, proposed electronic reporting and recordkeeping rule, EPA, 270-272
Aventis, *See* AgrEvo

B

Backup power for laptop batteries, 45-46
Backup system and procedures, 259-260
Bayer Corporation, electronic capture of field residue trial data, 34-39
Bayer-Pest Management Regulatory Agency collaboration, 217-219
Bayer residue programs, growth since 1994, 38-39
Bayer standard operating procedures in FieldNotes™, 48
Bid document preparation in Advantage™, 60-61
Biological and Pollution Prevention Division, EPA, 175
Business-critical systems, definition, 118
Business logic tier in Astrix agrochemical e-compliance architecture, 31

C

CADDY system
  adoption in Europe, 212-214
  electronic submission of regulatory dossiers, 209-215
  future development, 214-215
  objectives and functions, 211-212
  See also Joint Data Transfer Steering Group
Calculations, reports and graphs in detailed system design, 115
Calibration, equipment and systems, 105-106
Canadian pesticide regulatory agency, *See* Pest Management Regulatory Agency
CDX, *See* Central Data Exchange
Central Data Exchange, infrastructure supporting proposed electronic reporting and record-keeping rule, EPA
  architecture, 319-323
  building blocks, 314-327
  certifying submissions, uniformity, 320-321
  development, 248-249, 270-272, 312-313
  document submission, 284-286

registration process, 316-319
Certificate for individual in public key infrastructure, 315
Certificate issuance by Access Certificates for Electronic Services, 317-319
Certification with Respect to Data Integrity, 204
Certified limits and nominal concentration, 183-185
40 CFR, proposed part 3, Electronic Reporting: Electronic Records overview of regulation, 281-283t
text of rule, 334-346
Children
pesticide residue determination on foods commonly consumed by, 155-157
protection by FQPA, 152-153
Chromatography data acquisition system, system requirement specification, 76-77
Chromeleon^{TM} data capture system, 66-74
CICOPLAFEST, Zoxamide fungicide data submission, 220-221
Client-server architecture, Central Data Exchange, 321
Clinton administration role in protecting U.S. children, 152-153
Clinton-Gore administration, conducting government business electronically, 231
Closed system, definition, 245t-246
Committee role in data collection process, 254-255
Compliance critical systems, definition, 117-118
Compliance with U.S. environmental protection program based on self-reporting, 233
Computer Aided Dossier Design and Supply, See CADDY

Computer equipment, maintenance and repair records, 47-48
Computer validation umbrella, 2
Computerized systems in facility inspections, 83-84
Confidential Business Information
delayed electronic filing, 280
improper access, 204-205
Consolidation in agrochemical industry, 19-20
"Consultation and Coordination with Indian Tribal Governments" Executive Order 13175, 333
Copy of record with digital signatures, retention issues, electronic signature/certification scenario, proposed electronic reporting and recordkeeping rule, EPA, 305-306
Cost-benefit analysis, electronic report receiving systems, proposed electronic reporting and recordkeeping rule, EPA, 306-312
Cost-benefit analysis for implementation of FieldNotes^{TM}, 38
Cost-benefit considerations for validated compliant systems, 255-256
Cost savings in electronic data capture, 36
Cost to implement and operate electronic reporting and recordkeeping, proposed electronic reporting and record-keeping rule, EPA, 310-312
Costs, relative distribution across functional areas, 307-308
Coverage, Electronic Reporting: Electronic Records, proposed rule, [text], 335
CROMERRR (Cross-media Electronic Reporting and Recordkeeping Rule), 14, 87, 94, 232-238, 242-251t

353

See Also EPA proposed rule: Electronic Reporting: Electronic Records, [text], 334-346
See Also EPA proposed rule on electronic reporting and recordkeeping, official commentary, 264-333
Cross-Media Electronic Reporting and Record-keeping Rule, *See* CROMERRR
Cumulative risk assessment in FQPA, 152-157
Current pesticide tolerances reassessment, FQPA, 152

D

Data access control in Chromeleon^{TM} system, 71-74
Data and report audits by quality assurance unit, 82-83
Data backup system, requirements, 46
Data collection and storage, typical hardware configuration, 69*f*
Data collection justification by EPA, 233-234
Data collection process, committee role, 254-255
Data collection system choice, 67, 70
Data collection systems, 253-262
Data entry for computerized systems in clinical trials, 90-91
Data evaluation records, EPA and ACPA joint project, 141
Data exchange flexibility, Central Data Exchange, 320-321
Data integrity, FIFRA requirements, 99-100
Data retrieval in electronic record systems, 91-92
Data review online, Central Data Exchange, 326
Data revision record generation in Advantage^{TM}, 63
Data revision record, part of GLP compliance, 25
Data stored and manipulated, integrity, 249-250
Data transfer system, requirements, 46
Date/time stamps in electronic record systems, 91
Definitions, Electronic Reporting: Electronic Records, proposed rule, [text], 336-338
Design phase in software development life cycle, 120-121
Desktop computer in office, audit considerations, 44
Development considerations, Advantage^{TM} electronic Field Trial Notebook, 23-26
Development phase in software development life cycle, 121
Digital signatures, definition, 245*t*
See also Public key infrastructure
Document creation in PDF, 135-136
Document integrity, in electronic transmission, legal issue, 234
Documentation, equipment and systems, 107
Documentation requirements for software design, 109-116
Documents in PDF, benefits, 137
DOJ, electronic processes, 233-234
Drinking water, protocol for organophosphate pesticides monitoring, 154-155
DuPont, corporate records and information management, 126
Durability management in electronic data archiving, 128-129

E

E-compliance module in Astrix system, 32-33
E-government, legal and policy issues, 229-240
E-SIGN legislation, 231-232, 278-280
Ease of use in eFTN development, 23-24
Ecotoxicology in pesticide dossier, 222
EDI Implementation Guideline, 1996 Policy, EPA, 86, 95, 272-274
EDI/XML submissions to EPA, facility regulatory cost reduction estimates, 308-309t
Electronic authorization or signature, timing synchronization, 10-11
Electronic capture, GLP field data, reasons for increased use, 19-20
Electronic data archiving process, strategic components, 126-127
Electronic data archiving to ensure accessibility, durability and usability, 124-132
Electronic data capture methods
advantages and disadvantages, 36-37
increased use in agrochemical industry, 19-20
Electronic data collection, study director's viewpoint, 40-43
Electronic data harmonization, Pest Management Regulatory Agency, Canada, 217-218
Electronic data interchange (EDI) standards, Central Data Exchange, 323-325
Electronic Data Interchange, FIPS PUB 161-2, 325
Electronic data systems, GLP possible application, 250-251t
Electronic data submission benefits, 139-150
obstacles at state level, 161-163
pilot efforts at EPA, 188-197
success factors, 218
treatment of supplemental files, 198-208
Electronic data submission applications
government reviewers' needs, 146
verification, pilot submissions, 147
Electronic data submission for pesticides
historical perspective, 141-144
international harmonization, 148-150
pesticide registration 139-150
states for pesticide registration, 159-163
Electronic data submission for regulatory dossiers in Europe, 209-215
Electronic data submission to Pest Management Regulatory Agency
pilot program with international input, 216-219
Zoxamide fungicide registration, 223-228t
Electronic field trial notebook, 18-27
Electronic government, *See* E-government
Electronic information exchange and interaction organization, EPA and ACPA project, 142
Electronic label comparison, current EPA effort, 170-172
Electronic pesticide labeling, EPA, 164-173
Electronic pesticide submissions, efficiency, 133-138
Electronic record, definition, 89, 245t
Electronic record-retention systems, criteria, proposed electronic reporting and recordkeeping rule, EPA, 286-287

Electronic recordkeeping requirements in 21 CFR Part 11, 7-9
Electronic Recordkeeping under EPA-Approved State Programs, Electronic Reporting: Electronic Records, proposed rule, [text], 341-346
Electronic Recordkeeping under EPA Programs, Electronic Reporting: Electronic Records, proposed rule, [text], 339-341
Electronic records
  advantages to implementation, 14
  general GLP compliance, 2-3
  potential issues in field trials, 15
  problems for quality assurance people, 16
  retention requirements, 21 CFR 11, 7-9
  validation, 21 CFR 11, 87-88
Electronic Records and Electronic Signatures, FDA final rule, 3-4, 87-89, 242
Electronic records with electronic signatures, proposed electronic reporting and recordkeeping rule, EPA, 287-288
Electronic reporting and recordkeeping, proposed rule, EPA, effect, 269-272
"Electronic Reporting Benefit/Cost Justification Report" 1999, 307
Electronic Reporting: Electronic Records, EPA proposed rule, [text], 334-346
Electronic reporting, EPA 1996 policy, 272-274
Electronic signature/certification scenario, electronic report receiving systems, proposed electronic reporting and recordkeeping rule, EPA, 299-304
Electronic signature, definition, 89, 245t

Electronic signature method, criteria for electronic report receiving systems, proposed electronic reporting and recordkeeping rule, EPA, 293
Electronic signature requirements, predicate rule in 21 CFR Part 58, 7-9
Electronic Signatures in Global and National Commerce Act of 2000, *See* E-SIGN legislation
Electronic SOP systems, requirements according to FDA final rule, 8-9
Electronic SOPs, separation from electronic policies and guidelines, 10
Electronic Standards for the Transmission of Regulatory Information, FDA, 95
Electronically captured analytical chemistry data, audit process, 78-85
Electronically maintained records, requirements, proposed electronic reporting and recordkeeping rule, EPA, 286
Emergency situations, availability of paper records, 237
Encryption, automatic proprietary and compression process, 25
Encryption in security for data exchanges, Central Data Exchange, 323
End-use product, definition, 183
Endocrine effects testing in FQPA, 152-157
Enforceability of environmental programs, data integrity, 249-251t
Enforcement actions, criteria for electronic report receiving systems, 291-292
Entities affected by proposed electronic reporting rule, 265-266t
Environmental compliance reports, electronic reporting, 229-240

Environmental fate and effects, pilot electronic data submission to EPA, 207
Environmental fate in pesticide dossier, 222
Environmental Protection Agency, *See* EPA, specific topics
EPA and ACPA joint projects, 141-142, 195, 201
EPA cooperation with Pest Management Regulatory Agency, 142-143, 148, 195
EPA, electronic label comparison, 170-172
EPA, electronic records and electronic signatures, 242-244
EPA, electronic reporting and recordkeeping strategy, 232-234
EPA, electronic reporting policy, 272-276
EPA establishment of Electronic Reporting: Electronic Records, proposed rule, 264-346
EPA, *Filing of Electronic Reports via Electronic Data Interchange*, 86, 94-95
EPA–industry definitions for supplemental files, 201-205
EPA participation in the Joint Data Transfer Steering Group, 211
EPA pilot efforts for electronic data submission, 188-197
EPA proposed electronic reporting and recordkeeping rule, purpose, 268-272
EPA proposed rule: Electronic Reporting: Electronic Records, [text], 334-346
See Also CROMERRR
EPA proposed rule on electronic reporting and recordkeeping, official commentary, 264-333
See Also CROMERRR
EPA provisions for state delegated programs, 236-237
EPA reviewers, efficiency in pesticide registration, electronic data submission, 140-144
EPA Strategic Plan, 232
EPA, 10X-Safety Factor task force, 153
EPA tips to pesticide registrants/applicants, 186-187
EPA, Zoxamide fungicide, electronic data submission–pilot program, 220-228*t*
Ergonomic issues in adoption of CADDY, 213-214
Establishment of Electronic Reporting: Electronic Records, proposed rule, 264-346
ESTRI, *See Electronic Standards for the Transmission of Regulatory Information*
European Council Directive for regulation of plant protection products, 209-210
European Crop Protection Association, development of electronic dossier submissions, 210
European Union, electronic data submissions for regulatory dossiers, 209-215
Evaluation group, action plan, 256
Evaluator needs, Pest Management Regulatory Agency, 219
Executive Order 12866 on significant regulatory action, 327
Executive Order 13045 "Protection of Children from Environmental Health Risks and Safety Risks", 332
Executive Order 13132 "Federalism", 327-328
Executive Order 13175 "Consultation and Coordination with Indian Tribal Governments", 333

Executive Order 13211 "Energy Effects", 333
Extended mark-up language, *See* XML data exchange

F

Facility inspections including computerized systems, 83-84
FDA criteria, in proposed electronic reporting and recordkeeping rule, EPA, 288
FDA electronic data submission, application to pesticide registration, 143
FDA, *Electronic Standards for the Transmission of Regulatory Information*, 95
FDA, Final Rule, Electronic Records and Electronic Signatures, 3-4, 8-9, 87-89
FDA, *Guidance for Industry, Computerized Systems Used in Clinical Trials*, 87, 89-94
FDA, *Guidance for Industry, Providing Regulatory Submissions in Electronic Format–General Considerations*, 87, 95-97, 134-135
FDA signature requirements, 7-9
Federal Food, Drug, and Cosmetic Act, *See* FFDCA
Federal Implementation Convention, Central Data Exchange, 325
Federal Information Processing Standards on Internet, 316
Federal Insecticide, Fungicide, and Rodenticide Act, *See* FIFRA
"Federalism", Executive Order 13132, 327-328
FFDCA, as amended by FQPA, 175, 199
Field residue studies for pesticide registration, 35
Field residue trial data at Bayer Corporation, electronic data collection, 34-39
Field trial notebook generator in Advantage^{TM}, 59
Field trials, potential problems with electronic records, 15
FieldNotes^{TM}
  availability to quality assurance auditor, 50
  generic menu choices, 52-53
  handling of plot maps, 51
  test substance tracking, 52
  use by Bayer Corporation, 37-39
  use by AgrEvo Research Center, 41-43
FIFRA, as amended by Food Quality Protection Act of 1996, 152, 175, 199
FIFRA data captured electronically, 79-80
FIFRA regulations for field trials, 17
FIFRA, requirements for data integrity, 99-100
FIFRA requirements for pesticide "master label", 166
FIFRA/TSCA reporting by electronic means, 241-252
Filing of Electronic Reports via Electronic Data Interchange, EPA, 86, 94-95
Final study report in Advantage^{TM}, 60
Firewall security, Central Data Exchange, 321-322
Flexibility in exchanging data, Central Data Exchange, 320-321
Florida Department of Agriculture, pioneer in regulating pesticide residues, 41
Food and Drug Administration, *See* FDA
Food Quality Protection Act of 1996, *See* FQPA

Food supply assessment, organophosphate pesticides, 155-157
Form driven software program, Advantage™, 24
FQPA, 151-157, 199
Freedom of Information Act and supplemental file release, 204
Functional decomposition in detailed system design, 113
Functionality, Advantage™ Field Trial Manager, 20-22
Functionality development, in Advantage™, 62-63
Fungicide, Zoxamide, electronic data submission, 220-228t

G

General principles for computerized systems in clinical trials, 90
General Services Administration, Access Certificates for Electronic Services, 317
Generic menu choices suitability in audit, 52-53
Global features and integration in detailed system design, 114
GLP compliance
 by electronic records, general, 2-3
 in eFTN development, 24-26
GLP field data, reasons for increased use of electronic capture methods, 19-20
GLP regulatory citations applying to electronic data systems, 250-251t
Goals, proposed electronic reporting and recordkeeping rule, EPA, 276-277
Good Laboratory Practice Standards, *See* GLP
Government Paperwork Elimination Act of 1998, 4, 231, 328-329

Government reviewers' needs in electronic data submission applications, 146
Guidance documents for auditing electronically captured data, 80
Guidance documents, Pest Management Regulatory Agency, 218-219
Guidance for Industry, Computerized Systems Used in Clinical Trials, FDA, 87, 89-94
Guidance for Industry, Providing Regulatory Submissions in Electronic Format–General Considerations, FDA, 87, 95-97

H

Hardware configuration for data collection and storage, 69f
Harmonized Test Guidelines publication on Internet, EPA, 199
Hazardous Waste Manifest, electronic reporting rule, 280, 307
Health Effects Division, EPA, 175
Historical perspective, electronic data submission for pesticide registration, 141-144
Historical SOP versions, electronically archived, 7
HPLC system control from a workstation, 68f
Hybrid system, mistake in standard operating procedures coordination, 6
Hybrid system proposal with both electronic and paper records, 16-17

I

Identity proofing by certificate authority, 317-318

Implementation conventions, transaction set, Central Data Exchange, 324-325
Implementation, Electronic Reporting: Electronic Records, proposed rule, [text], 335-336
Improvement in pesticide registration process, 144-146
Improvements in pesticide labeling submissions using electronic tools, 169-170
In-process inspections by quality assurance auditor, 81-82
Indian tribal governments, consultation and coordination, Executive Order 13175, 333
Industry-government partnerships, 147-148
Industry-Pest Management Regulatory Agency, collaboration, 217-219
Industry role in FQPA assessment methodology, 154-157
Industry self-reporting, environmental compliance, 233
Initiation phase in software development life cycle, 119
Installation phase in software development life cycle, 122-123
Integrity, data stored and manipulated, 249-250
"Intent to be bound" in electronic transmission, legal issue, 235
International harmonization, electronic data submission, pesticides, 148-150
Internet, *Agency Harmonized Test Guidelines*, EPA, 199
Internet availability of pesticide labels, 169
Internet site for Office of Pesticide Programs pilot efforts, 195, 207-208
Internet, state pesticide registration data, 163
Inventory management in metrology program, 101-102*f*

J

Joint Data Transfer Steering Group, 211
See also CADDY

K

Kelly Registration Systems, developer of state pesticide registration software, 160-161, 163

L

Label comparison, current EPA effort, 170-172
Label registration process (paper based) for pesticides, 167-169
Labeling, pesticides by electronic means, 164-173
Labels, accepted pesticide, availability to interested parties on Internet, 169
Laboratory technology in life cycle management, 127
Laptop computer in field, audit considerations, 44-46
Late data entry, requirement for explanation, 52
Legal and policy issues, e-government, 229-240
Life cycle, instrument or system, 101, 103*f*-105
Limit of Detection for method, 261
Limit of Quantitation for method, 261
Logical application design in detailed system design, 114
Logical data design, software development, 112-113
Logical security in electronic data systems, 92

M

Machiavelli, quotation from *The Prince* on a new order of things, 134
Magnetic media exclusion in proposed electronic reporting and recordkeeping rule, EPA, 270
Maintenance, equipment and systems, 105-106
Maintenance records for computer equipment, 47-48
Manufacturing-use product, definition, 183
"Master label", FIFRA requirements for pesticides, 166
Maximum Likelihood Imputation Procedure, organophosphate pesticide residue calculations, 155
Metrology program elements, 100f-101
Metrology, science of measurement, 98-108
Mexican environmental regulatory agency, *See* CICOPLAFEST
Millennium^{32} chromatography software validation, 75-78
Minimum effective rate determination for pesticides, 35
Missing data in residue field reports, 39
Mistakes in managing electronic standard operating procedures, 5-11
Multisite uses, electronic SOPs, protocols and amendments, 9-10

N

NAFTA pesticide review, Zoxamide fungicide, 220-221
NAFTA Technical Working Group, electronic data submission, 148
National Agricultural Chemicals Association, *See* American Crop Protection Association
National Archives and Records Administration, paper and microfiche policy, 196-197
National Technology Transfer and Advancement Act, 331-332
Networks, computer system and components, 258-259
Nominal concentration and certified limits, 183-186
Non-repudiation in electronic transmission, legal issue, 235
Notice of Agency's General Policy for Accepting Filing of Environmental Reports via Electronic Data Interchange, 1996 policy, 272-274

O

Occupational-residential pilot electronic data submission to EPA, 206
OECD dossier submission for Zoxamide fungicide, 221
OECD harmonization with EPA test procedures, 175
OECD Working Group on Pesticides, 148
Official commentary, proposed electronic reporting and recordkeeping rule, EPA, 264-333
Office of Environmental Information, 14, 231
Office of Pollution Prevention and Toxics, *See* OPPTS
Open system, definition, 245t-246
Operation qualification, system, 104
Operations phase in software development life cycle, 123
OPPTS test guidelines and proposed upgrades, 176-183

361

Organophophate Market Basket Survey Task Force, 155-157
Organophosphate pesticides
  food supply, assessment, 155-157
  OP Case Study Group, 154-157
  protocol for drinking water monitoring, 154-155

P

Paper based label registration process for pesticides, 167-169
Paper records more reliable than electronic records, 21 CFR Part 11, 16
Paper vs.electronic data audit, 49-50
Paperwork Reduction Act
  mandate to streamline regulatory processes, 230, 269
  proposed electronic reporting and recordkeeping rule, EPA, 328-330
Payables module in Advantage^{TM}, 61
PDF documents
  benefits, 137, 226
  by using Adobe Acrobat software, 135-136, 191-194, 196
Performance qualification, system, 104-105
Personal information selected for identify proofing access, 318
Personnel training for computerized system, 93
Pest Management Regulatory Agency
  electronic submission, 216-228*t*
  guidance documents, 218-219
  Joint Data Transfer Steering Group, 211
  pesticide registration data harmonization, 142-143, 148
  Zoxamide fungicide, electronic data submission, 220-228*t*
Pesticide Data Program, USDA, 155
Pesticide labeling
  basic requirements, 165-167
  current label registration process (paper based), 167-169
  electronic, 164-173
  submissions improvements using electronic tools, 169-170
Pesticide labels, availability to interested parties on Internet, 169
Pesticide registrants/applicants, EPA tips, 186-187
Pesticide registration
  data on Internet, 163
  electronic data submission, 139-150
  electronic data submission to states, 159-163
  FDA electronic data submission rules, possible application, 143
  requirement for field residue studies, 35
  submission, components, 221-222
Pesticide registration process, improvements, 144-146
Pesticide registration, supplemental files issues and concerns
  archiving, 204
  certification of authenticity, 204
  definition, 201-202
  formatting, 203
  over-analysis, 203
  security, 204-205
Pesticide regulatory program, EPA, study data significance, 190-191
Pesticide residue determination on foods commonly consumed by children, 155-157
Pesticides
  efficiency in electronic submissions, 133-138
  international harmonization for electronic data submission, 148-150
  minimum effective rate determination, 35

Physical security in electronic data systems, 92
Physicochemical properties in pesticide dossier, 221-222
Pilot efforts for electronic data submission at EPA, 188-197
Pilot submissions in electronic data applications verification, 147
Pioneer in regulating pesticide residues, Florida Department of Agriculture, 41
PKI, *See* Public key infrastructure
Plant protection products regulation, European Council Directive, 209-210
Plot maps in electronic data audit, 51
Policies and guidelines, electronic, separation from electronic SOPs, 10
Portable Document Format, *See* PDF
Predicate rule for electronic signature requirements in 21 CFR Part 58, 7-9
Private key creation, Central Data Exchange, 323
Product chemistry data requirements, pesticide chemistry registration, 174-187
Product management assistance-database in Advantage™, 60
Program interface architecture in detailed system design, 114
Project assessment and development for software, overview, 110-111*f*
Project plan in detailed system design, 115
Proposed rule, Electronic Reporting: Electronic Records, [text], 334-346
Proposed rule on establishment of electronic reporting: electronic records, EPA, 264-346
"Protection of Children from Environmental Health Risks and Safety Risks", Executive Order 13045, 332
Protocol generator in Advantage™, 58

Public access to environmental compliance information, proposed electronic reporting and recordkeeping rule, EPA, effect, 269-270
Public key infrastructure, 249, 275, 305
See also Digital signatures
Public key infrastructure-based digital signatures in Central Data Exchange, 314-316
Published literature supporting electronic SOPs, availability and currency, 7
Punch cards, Sandoz Crop Protection analytical data submission, 41
Pure active ingredient, definition, 183
Purpose, proposed electronic reporting and recordkeeping rule, EPA, 268-272

Q

Qualitative implications, cost-benefit analysis, electronic report receiving systems, proposed electronic reporting and recordkeeping rule, EPA, 306-312
Quality assurance
 checklist for electronic records audit, 49
 independent check of calculations, 50
Quality assurance auditor, in-process inspections, 81-82
Quality assurance plan in detailed system design, 115
Quality assurance unit
 data and report audits, 82-83
 problems with electronic records, 16
 procedures for storing and archiving electronic data, 83
 training for personnel, 80-81

Quality in Advantage™, 62-63
Questions about data collection process, 254-257

R

Record-retention systems, availability in emergency situations, 237
Records inspection in computerized system, 94
Records management, basic definitions in 36 CFR, 125-126
Registration agreement in Central Data Exchange, 319
Registration Division, EPA, 175
Registration process, Central Data Exchange, 316-319
Regulation of plant protection products, European Council Directive, 209-210
Regulatory data requirements, product chemistry, 176-180
Regulatory Flexibility Act, analysis requirement, 330
Regulatory requirements for proposed electronic reporting and recordkeeping rule, EPA, 327-333
Reinventing Environmental Information Report, March, 1996, 269
Renewal, signature agreement, criteria for electronic report receiving systems, proposed electronic reporting and recordkeeping rule, EPA, 294-297
Repair records for computer equipment, 47-48
Report formats, EPA and ACPA cooperation, 141
Requirement for user friendly system, 11

Requirements definition phase, software development, 110, 112, 119-120
RESIDAT, database used by AgrEvo for electronic data submission for pesticide registration, 42
Residue chemistry, pilot electronic data submission to EPA, 206
Residue studies in pesticide dossier, 222
Responses from electronic report receiving systems in electronic signature/certification scenario, proposed electronic reporting and recordkeeping rule, EPA, 302-304
Risk assessments in pesticide dossier, 222
Rohm and Haas, Zoxamide fungicide pesticide registration, 220-228t

S

Safety margins, increased by FQPA, 152-157
Sample tracking in Advantage™, 58-59
Sandoz Crop Protection, analytical data submission on punched cards, 41
Scope and method, cost-benefit analysis, electronic report receiving systems, proposed electronic reporting and recordkeeping rule, EPA, 306-312
Security, hardware and software audit, electronic field data, 46
Signature agreement, criteria for electronic report receiving systems, proposed electronic reporting and recordkeeping rule, EPA, 294-295
Signature agreement, renewal, criteria for electronic report receiving systems, proposed electronic

reporting and recordkeeping rule, EPA, 295-297
Signature and certification scenario in transaction environment, proposed electronic reporting and recordkeeping rule, EPA, 325-327
Significant regulatory action under Executive Order 12866, 327
Small business, effect of proposed electronic reporting and recordkeeping rule, EPA, 330
Small governments, State, local, and tribal, effect of proposed electronic reporting and recordkeeping rule, EPA, 330-331
Software design, documentation requirements, 109-116
Software development life cycle, validation importance, 118-123
Software project assessment and development, overview, 110-111f
Software validation in Advantage^{TM}, 63
SOPs, *See* Standard Operating Procedures
Special Review and Reregistration Division, EPA, 175
Stakeholders consultation, proposed electronic reporting and recordkeeping rule, EPA, 277-278
Standard components for acceptable validation, electronic SOP, protocol and amendment systems, 9
Standard Operating Procedures
  computerized systems in clinical trials, 90
  mistakes in managing electronic, 5-11
  required for electronic data, 48-49
Standardized residue chemistry study report, 141
Standards, methods, tools and errors in detailed system design, 114

State business needs, provisions for delegated programs, proposed electronic reporting and record-keeping rule, EPA, 236-237
State Electronic Commerce/Electronic Data Interchange Steering Committee, 277
State Guide for Electronic Reporting of Environmental Data, 277
State, local and tribal programs, criteria for electronic report receiving systems, proposed electronic reporting and recordkeeping rule, EPA, 291-306
State, local and tribal programs, provision for approval changes, proposed electronic reporting and recordkeeping rule, EPA, 289-290
State, local and tribal programs, provision for electronic reporting, proposed electronic reporting and recordkeeping rule, EPA, 270-272
State obstacles to electronic data submission, 161-163
State pesticide administrator, 160
State pesticide registration software by Kelly Registration Systems, 160-161
State pesticide registrations, electronic data submission, 159-163
Storage media issues, proposed electronic reporting and recordkeeping rule, EPA, 247, 288
Storing and archiving electronic data, review procedures by quality assurance unit, 83
Strategic components in electronic data archiving process, 126-127
Study data significance in EPA pesticide regulatory program, 190-191
Study director
  data access control, 73-74
  management functions, 55t

viewpoint on electronic data collection, 40-43
Study management efficiency maximization in Advantage™, 61-62
Study reconstruction in electronic data systems, 92
Submitter registration process, criteria for electronic report receiving systems, proposed electronic reporting and recordkeeping rule, EPA, 291-306
Success factors, electronic data submission, 218
Supplemental files
  definition in pesticide registration, 201-202
  electronic data submission to EPA, 198-208
  issues and concerns in pesticide registration, 203-205
Surrender certification, electronic report receiving systems, proposed electronic reporting and recordkeeping rule, EPA, 297-298
System administration in detailed system design, 114
System administrator's role in data access control, 72-73, 260-261
System archives, electronic signature/certification scenario, electronic report receiving systems, proposed electronic reporting and recordkeeping rule, EPA, 304-306
System controls in computerized system, 93
System dependability in computerized system, 92-93
System description and protocol requirements, 258-262
System design, specification documentation, 113-115
System qualification in Chromeleon™, 71

System requirement specification, chromatography data acquisition system, 76-77
System security
  Advantage™ eFTN, 24-26
  Central Data Exchange, 320-323
  model in detailed system design, 114
System security requirements, criteria for electronic report receiving systems, proposed electronic reporting and recordkeeping rule, EPA, 292-293
System validation documentation, multisite uses, 10
System validation in quality assurance process, 47, 261
System test plan in detailed system design, 115

T

Technical architecture phase, software development, 112
Technical grade of active ingredient, definition, 182-183
Technology-neutral criteria for electronic reporting, EPA, 235-236
Technology transitions in successful electronic data archiving processes, 126-127
10X-Safety Factor task force, EPA, 153
Test phase in software development life cycle, 121-122
Test plan preparation by application administrator, 77-78
Test substance tracking, 52
Timing, electronic authorization or signature, 10-11
Toxic Substances Control Act, *See* TSCA

Toxicology data format, EPA and ACPA cooperation, 141-142
Toxicology in pesticide dossier, 222
Toxicology, pilot electronic data submission to EPA, 206-207
Training recommendation for system users, 262
Transaction environment, Central Data Exchange, 325-327
Transaction record, electronic signature/certification scenario, electronic report receiving systems, proposed electronic reporting and recordkeeping rule, EPA, 304
TSCA/FIFRA reporting by electronic means, 241-252
Two-person requirement in system administration, 260

U

U.S. Department of Justice, *See* DOJ
U.S. Environmental Protection Agency, *See* EPA
U.S. Food and Drug Administration, *See* FDA
Unfunded Mandates Reform Act, effect on proposed electronic reporting and recordkeeping rule, EPA, 330-331
Uniformity in certifying submissions, Central Data Exchange, 320-321
Usability management in electronic data archiving, 129-130
User acceptance test plan for chromatography data acquisition system, 76-77
User economics in eFTN development, 23
User friendly system requirement, 11
User interface in Astrix agrochemical e-compliance architecture, 31

User privileges and access to system, 259-260
User testing strategy for software validation, 75-78
User validation in eFTN development, 26
User verification in quality assurance process, 47

V

Validated compliant systems, cost-benefit considerations, 255-256
Validation
 definition, 109-110
 electronic SOP, protocol and amendment systems, 9
 Millennium^{32} chromatography software, 75-78
 system data and software, 261
Validation in a regulatory environment, 117-123
Validation in software development life cycle, 118-123
Validity of data, criteria for electronic report receiving systems, proposed electronic reporting and recordkeeping rule, EPA, 291-293
Vendor audit, Astrix by Bayer Corporation, 37-38
Vendor responsibility in validation, system data collection and processing, 261
Voluntary consensus standards use in proposed electronic reporting and recordkeeping rule, EPA, 331-332

W

Warning statements, electronic signature/certification scenario, electronic report receiving

systems, proposed electronic reporting and recordkeeping rule, 300-302
Web-based Advantage™ software, 64-65
Web-based submissions to EPA, facility regulatory cost reduction estimates, 308-309*t*
Web-centric computer systems by Astrix Software Technology, 28-33
Web forms, standards based, 331-332

X

XML data exchange format, 324, 332

Z

Zoxamide fungicide, electronic data submissions, 220-228*t*